T0262933

CRC SERIES IN ENZYME BIOLOGY

Phosphatidate Phosphohydrolase

Volume I

Editor

David N. Brindley, Ph.D., D.Sc.

Heritage Medical Scientist
Professor of Biochemistry
Lipid and Lipoprotein Group
Faculty of Medicine
University of Alberta
Edmonton, Canada

CRC Series in Enzyme Biology

Series Editor-in-Chief
John R. Sabine, Ph.D.

CRC Press
Taylor & Francis Group
Boca Raton London New York

CRC Press is an imprint of the
Taylor & Francis Group, an **informa** business

First published 1988 by CRC Press
Taylor & Francis Group
6000 Broken Sound Parkway NW, Suite 300
Boca Raton, FL 33487-2742

Reissued 2018 by CRC Press

© 1988 by Taylor & Francis
CRC Press is an imprint of Taylor & Francis Group, an Informa business

No claim to original U.S. Government works

A Library of Congress record exists under LC control number: 87020867

Publisher's Note
The publisher has gone to great lengths to ensure the quality of this reprint but points out that some imperfections in the original copies may be apparent.

Disclaimer
The publisher has made every effort to trace copyright holders and welcomes correspondence from those they have been unable to contact.

ISBN 13: 978-1-138-50573-5 (hbk)
ISBN 13: 978-1-138-56117-5 (pbk)
ISBN 13: 978-0-203-71106-4 (ebk)

Visit the Taylor & Francis Web site at http://www.taylorandfrancis.com and the
CRC Press Web site at http://www.crcpress.com

SERIES PREFACE

The "CRC Series on Enzyme Biology" is a series of books, each one devoted to a single enzyme and each one endeavoring to draw together in a comprehensive and systematic manner all that is currently known about the chemistry, biochemistry, and physiology of that particular enzyme. Each volume, written or edited by one or more international enzyme specialists, draws together the latest information available on the occurrence, structure, function, role, and control of the biologically more important enzymes.

Each chapter or section of each volume is not written primarily for those specializing in that narrow area (innumerable specialist reviews do this), but rather to provide a coherent and integrated summary of that aspect for those working on other, quite different facets of the same enzyme. In this way several distinct classes of scientist, and student, will benefit from each volume, namely, those concerned with "type" situations of which that enzyme is an example, those working on the metabolic systems of which that enzyme is a key component, and those concerned with diseases in which the enzyme has an important role.

PREFACE

Phosphatidate phosphohydrolase is an enzyme that catalyzes a critical reaction in the synthesis of glycerolipids. The diacylglycerol that it produces is the precursor for the synthesis of triacylglycerols, phosphatidycholine, and phosphatidylethanolamine, and in plants for galactolipids. Consequently, its activity is essential in most tissues especially in order to provide the phospholipids that are needed for membrane formation. Because of this, no attempt has been made to review all papers on phosphatidate phosphohydrolase but rather we have selected tissues in which the enzyme has been characterized fairly extensively and for which there is a reasonable body of evidence concerning its role in metabolic regulation.

The mammalian tissues that have been chosen are liver, lung, and adipose tissue since their requirements for glycerolipid synthesis are specialized and fairly different. There is a further chapter that deals with plants and microorganisms. Each of these chapters has a general section that describes the special needs for glycerolipid synthesis and the physiological context in which the regulation of phosphatidate phosphohydrolase activity can be understood.

On a personal level, I should like to thank all of my colleagues who have worked with me on phosphatidate phosphohydrolase. I particularly wish to mention those who have offered criticism and advice in the preparation of these volumes and those whose work is illustrated in Chapters 1 and 2. These include: Mariana Bowley, Paul Bracken, Sue Burditt, Carmen Cascales, Maria Cascales, June Cooling, Robin Fears, Paul Hales, Roger Hopewell, Helen Glenny, Katherine Lloyd-Davies, Heather Mangiapane, Ashley Martin, Paloma Martin-Sanz, Janette Morgan, Sylva Pawson, Richard Pittner, Haydn Pritchard, Andy Salter, Janice Saxton, and Graham Sturton.

Finally, my deep gratitude to my Ph.D. supervisor and friend, the late Professor Georg Hubscher. He introduced me to phosphatidate phosphohydrolase and taught me so much about biochemistry.

SERIES EDITOR

John R. Sabine, M. Agr. Sc., Ph.D. is a Reader in Animal Physiology at the Waite Agricultural Research Institute of the University of Adelaide, Adelaide, Australia. Dr. Sabine obtained his Bachelor's and Master's degree in Agricultural Science from the University of Melbourne and then his Ph.D. in Animal Nutrition in the laboratory of Dr. B. Connor Johnson at the University of Illinois, Urbana. After several research appointments — Monash University, Australian National University, and the University of California, Berkeley, he was appointed to the faculty of University of Adelaide in 1967. At various times since then he has held Visiting Professorships at Brandeis University (Graduate Department of Biochemistry), the University of Stockholm (Wenner-Gren Institute), the University of Oklahoma Health Sciences Center (Department of Biochemistry and Molecular Biology), and the University of Kuwait (Department of Biochemistry), as well as Visiting Scholar appointments at Oxford University (Department of Clinical Medicine) and Harvard University (Chemistry Department).

Dr. Sabine is a member of several scientific societies, including particularly the Australian Biochemical Society, the Nutrition Society of Australia, and the Australian Physiological and Pharmacological Society. He has presented his research findings at various national and international meetings, and has been chairman of a number of sessions at these meetings. He was co-convenor of, and leader of the Australian delegation to, the U.S./Japan/Australia Cancer Conference (Hawaii, 1975) and the convenor and chairman of the unique international symposium "Lipids in Cancer", which was held on board the Indian-Pacific Express as a satellite meeting to the 12th International Congress of Biochemistry, 1982. In 1979 he delivered the 12th Patricia Chomley Oration to the Australian College of Nursing.

Dr. Sabine has published some 50 research articles and 14 invited reviews. He is on the Editorial Board for *Nutrition and Cancer*, and his book *Cholesterol* was the first comprehensive coverage of this important field to appear in 20 years. His major research interests revolve around cholesterol physiology, with particular reference to the control of its synthesis and to its role in membrane structure and function, and with emphasis upon the role of cholesterol in the etiology of cancer. He has further research interests in such diverse fields as the role of earthworms in biological resource recovery, the physiology of goats for meat and fiber production and the interaction between scientists and society.

THE EDITOR

David N. Brindley, Ph.D., D.Sc. was Professor of Metabolic Control in the Department of Biochemistry at the University of Nottingham, Nottingham, England.

Professor Brindley received his undergraduate training in the Department of Medical Biochemistry, University of Birmingham and received a B.Sc. (1st Class Honours) in 1963. He then studied for a Ph.D. in the same Department under the supervision of the late Professor G. Hübscher and gained his Ph.D. in 1966. A further year was then spent as a Postdoctoral Fellow with Professor Hübscher. The work was concerned with the control of glycerolipid synthesis in the small intestine in relation to fat absorption.

In 1967 Professor Brindley moved to Harvard University where he worked for two years as a Postdoctoral Fellow for Professor K. Bloch. The work investigated the synthesis of fatty acids in *Mycobacterium phlei*. A multienzyme complex of fatty acid synthetase was shown to occur in a bacterium for the first time and novel stimulating factors for fatty acid synthesis were demonstrated.

Professor Brindley retruned to England in 1969 to join the newly established Medical School in Nottingham. He was subsequently promoted from Lecturer to Senior Lecturer, Reader, and then Professor and he gained his D.Sc. from the University of Birmingham in 1977. From January 1, 1988, he has accepted the position of Heritage Medical Scientist and Professor of Biochemistry in the newly formed Lipid and Lipoprotein Research Group that is sponsored by the Alberta Heritage Foundation for Medical Research.

His main interests are the effects of hormones, metabolites, drugs, and diet in regulating: (a) glycerolipid synthesis particularly at the level of phosphatidate phosphohydrolase; (b) the secretion of very low density lipoprotein and lysophosphatidycholine from the liver; (c) the binding and uptake of low and high density lipoproteins by the liver; (d) insulin responsiveness and metabolism in adipose tissue; and (e) food intake and the level of circulating glucose and triacylglycerol.

CONTRIBUTORS

David N. Brindley, Ph.D., D.Sc.*
Heritage Medical Scientist
Professor of Biochemistry
Lipid and Lipoprotein Group
Faculty of Medicine
University of Alberta
Edmonton, Canada

John L. Harwood, Ph.D., D.Sc.
Professor
Department of Biochemistry
University College
Cardiff, Wales

Fred Possmayer, Ph.D.
Professor
Department of Obstetrics and Gynecology,
 and Biochemistry
University of Western Ontario
London, Ontario, Canada

Molly J. Price-Jones, Ph.D.
Research
Department of Biochemistry
University College
Cardiff, Wales

E. D. Saggerson, Ph.D., D.Sc.
Reader in Biochemistry
Biochemistry Department
University College London
London, England

* At the time these volumes were written, Dr. Brindley was Professor of Metabolic Control, Department of Biochemistry, University of Nottingham, Nottingham, England.

TABLE OF CONTENTS

Volume I

Volume II

Chapter 1

GENERAL INTRODUCTION

David N. Brindley

TABLE OF CONTENTS

I. INTRODUCTION AND PATHWAYS OF GLYCEROLIPID SYNTHESIS

Phosphatidate lies at an important branch point in metabolism in mammalian tissues. It can be converted to CDP-diacylglycerol which acts as a precursor for the synthesis of acidic phospholipids such as phosphatidylinositol, phosphatidylglycerol, and diphosphatidylglycerol (Figure 1). Alternatively, the phosphatidate can be converted to diacylglycerol through the action of phosphatidate phosphohydrolase (EC 3.1.3.4) which is the subject of this book. Diacylglycerol can then serve as a substrate for the synthesis of the zwitterionic phospholipids, phosphatidylcholine and phosphatidylethanolamine, or it can be converted into triacylglycerol. In plants diacylglycerol is also converted into galactolipids (Chapter 4, Figure 1). Chapter 4 also describes the role of phosphatidate phosphohydrolase activity in microorganisms. A further route of metabolism for phosphatidate in mammalian liver is its deacylation back to glycerol phosphate by phospholipase A type activities (Chapter 2, Sections II and V). This reaction forms part of a substrate cycle that can decrease the accumulation of phosphatidate in membranes and prevent it from being incorporated into glycerolipids.

Phosphatidate can be produced *de novo* from either glycerol phosphate or dihydroxyacetone phosphate. There appears to be at least three different acyltransferases that initiate this synthesis.[1] The first is a glycerol phosphate acyltransferase that is located in the endoplasmic reticulum on the cytosolic surface.[2] This can also esterify dihydroxyacetone phosphate and the two acyl acceptors are mutually competitive.[3-5] A second glycerol phosphate acyltransferase is located on the outer mitochondrial membrane.[6-8] This enzyme is susceptible to proteolytic degradation from either the inner[9,10] or cytoplasmic[10] surface. These results indicate that the enzyme is a transmembrane protein. However, these studies do not clearly identify the location of the active site. The mitochondrial acyltransferase will not accept dihydroxyacetone phosphate as a substrate nor is this compound a competitive inhibitor.[3,11] A third acyltransferase is found mainly in peroxisomes and this enzyme esterifies dihydroxyacetone phosphate and not glycerol phosphate.[12,13]

The separate identities of these three acyltransferases are confirmed by the observations that the acyltransferase in the endoplasmic reticulum is relatively sensitive to inhibition of sulfhydryl-group reagents,[13-17] proteolytic enzymes,[18] and heat.[17] Conversely, the mitochondrial glycerol phosphate acyltransferase is fairly resistant to these treatments[1,14-18] and the dihydroxyacetone phosphate acyltransferase activity in peroxisomal fractions can be stimulated by *N*-ethylmaleimide.[13,19,20] The mitochondrial glycerol phosphate acyltransferase also had a lower K_m for glycerol phosphate[17] and acyl-CoA esters[13,14] than the acyltransferase in the endoplasmic reticulum and it also shows a preference for long-chain saturated rather than unsaturated acyl-CoA esters.[6,14-16,21-22] The microsomal acyltransferase can use a variety of saturated and unsaturated acyl-CoA esters as substrates.[15,23,24]

The next reaction in the synthesis of phosphatidate from glycerol phosphate is the acylation of lysophosphatidate. This reaction is catalyzed by an acyltransferase that is different from the enzymes responsible for the initial acylation of glycerol phosphate.[22,25-27] The activity of lysophosphatidate acyltransferase is normally relatively low in mitochondrial compared with microsomal fractions.[28]

When acyldihydroxyacetone phosphate is synthesized in peroxisomes or in the endoplasmic reticulum it must be converted to lysophosphatidate by a reductase that uses NADPH before it can be further acylated to phosphatidate. Peroxisomes do not appear to possess lysophosphatidate acyltransferase or the other enzymes required for the synthesis of triacylglycerol or phosphatidylcholine.[29] Therefore, it seems likely that the acyldihydroxyacetone phosphate or lysophosphatidate must be transferred to the endoplasmic reticulum if it is to be converted into these end products.

Potentially, phosphatidate phosphohydrolase could act upon phosphatidate that is produced *de novo* in mitochondria or in the endoplasmic reticulum. The mitochondrial site seems less

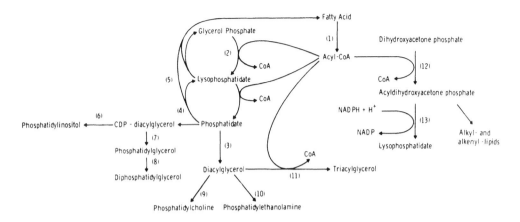

FIGURE 1. Pathways of glycerolipid biosynthesis. Enzyme activities are indicated by: (1) acyl-CoA synthetase (EC 6.2.1.3); (2) glycerol phosphate acyltransferase (EC 2.3.1.15); (3) phosphatidate phosphohydrolase (EC 3.1.3.4); (4) phosphatidate cytidylyltransferase (EC 2.7.7.41); (5) phospholipase A type activities; (6) CDP diacylglycerol-inositol 3-phosphatidyltransferase (EC 2.7.8.11); (7) glycerol phosphate phosphatidyltransferase (EC 2.7.8.5); (8) phosphatidyl transfer from CDP-diacylglycerol to phosphatidylglycerol; (9) choline phosphotransferase (EC 2.7.8.2); (10) ethanolamine phosphotransferase (EC 2.7.8.1); (11) diacylglycerol acyltransferase (EC 2.3.1.20); (12) dihydroxyacetone phosphate acyltransferase (EC 2.3.1.42); and (13) acyldihydroxyacetone phosphate reductase.

likely since these organelles generally appear to lack or have very low activities of the enzymes that can convert the diacylglycerol into triacylglycerol or phosphatidylcholine.[30,31] Mitochondria can, however, convert phosphatidate into CDP-diacylglycerol and use this to produce diphosphatidylglycerol (Figure 1). This lipid is an essential component of the inner mitochondrial membrane. In the livers of rat, guinea pigs, rabbits, and cows about 50% of the glycerol phosphate acyltransferase activity is mitochondrial[16,17,28] and this enzyme has a lower K_m value for acyl-CoA esters and glycerol phosphate compared to the enzyme in the endoplasmic reticulum.[13,15,17] This ought to imply that there should be a preferential esterification of fatty acids in mitochondria in the liver which ought to exceed the requirement for the synthesis of acidic phospholipids. It therefore remains to be established whether the fatty acids that are esterified as phosphatidate in mitochondria in liver are transferred to the endoplasmic reticulum and whether phosphatidate phosphohydrolase acts while the phosphatidate is still bound to mitochondria. In other tissues the activity of glycerol phosphate acyltransferase in mitochondria is relatively low compared to that in the endoplasmic reticulum.[1]

A further uncertainty in this area of metabolism is the extent to which phosphatidate is produced by the esterification of dihydroxyacetone phosphate rather than glycerol phosphate. It has been estimated that the acylation of glycerol phosphate should be about 84 times greater than that of dihydroxyacetone phosphate in the liver in vivo.[5] These conclusions and similar ones for adipose tissue[4,5] are based upon the kinetic parameters of the microsomal acyltransferase activities and the assumption that these values can be extrapolated to physiological situations. However, in these studies account was not taken of the specific glycerol phosphate and dihydroxyacetone phosphate acyltransferase activities that are found in mitochondria and peroxisomes respectively. These latter two activities have lower apparent K_m values for acyl-CoA esters than that of the acyltransferase of the endoplasmic reticulum.[13] Therefore, the former enzymes could preferentially esterify fatty acids when their concentrations are low.

Pollock et al.[33] measured the direct incorporation of dihydroxyacetone phosphate into phosphatidate by homogenates from several tissues and compared this rate with that for glycerol phosphate. Although the activities of glycerol 3-phosphate dehydrogenase (NAD^+) and glycerol phosphate acyltransferase always exceeded that of dihydroxyacetone phosphate acyltransferase this potential was not expressed and there was a predominant synthesis of glycerolipids from dihydroxyacetone phosphate.[33] When equimolar concentrations of glycerol phosphate and dihydroxycetone phosphate were incubated with "mitochondrial" fractions from rat liver[11] or with fractions of rabbit lung[34] there was an approximately equal rate of esterification with both of these precursors. Agranoff and Hajra[35] also concluded that the dihydroxyacetone phosphate pathway plays a significant role in glycerolipid synthesis in homogenates of mouse liver and a dominant role in homogenates of Ehrlich ascites tumor cells. These experiments were based upon the relative rates of incorporation of tritium from NAD[^{3}H] and NADP[^{3}H] into the C_2 position of glycerolipids. By contrast Declercq et al.[36] using near physiological concentrations of glycerol phosphate and dihydroxyacetone phosphate concluded that the combined glycerol phosphate acyltransferases contributed 93% of the glycerolipid synthesized by rat liver homogenates. It is possible that a contribution of dihydroxyacetone phosphate acyltransferase in peroxisomes might have been underestimated since this activity can be stimulated about 30-fold by pyrophosphate.[37] The reason for this is not clear at present. It may have resulted from a specific stimulating effect of pyrophosphate, or alternatively a nonspecific effect such as the chelation of polyvalent cations might have been responsible. It is also not known whether this effect of pyrophosphate has a physiological significance.

The experiments that have been described above are concerned with cell-free systems and further work has been performed to try to estimate the relative importance of glycerol phosphate and dihydroxyacetone phosphate as direct precursors for glycerolipid synthesis. All of these approaches have disadvantages which include problems of isotope effects, there being different pools of precursors in the cells, the choice of substrate, and the specificity of reduced pyridine nucleotides. Furthermore, the techniques are fairly complicated and some of them cannot be satisfactorily applied in vivo. Although this type of work enables us to make estimates of the potential of the two pathways for glycerolipid synthesis there is still no detailed information concerning how these pathways may change in importance in different physiological conditions. Further discussion of these problems can be found in Reference 1 and in Volume II, Chapter 5, Section XII.

Experiments with specifically labeled glycerol have shown that the dihydroxyacetone phosphate pathway could account for 50 to 75% of glycerolipid synthesis in rat liver slices[38,39] and 56 and 64% of the phosphatidylglycerol and phosphatidylcholine respectively that is synthesized by type II alveolar cells.[40] Other work with specifically labeled glucose molecules demonstrated that between 49 and 61% of glycerolipid synthesized by BHK-21 and BHK-Ts-a/lb-2 cells were formed by direct esterification of dihydroxyacetone phosphate.[41,42] Apart from its possible function in the synthesis of acylglycerolipids the esterification of dihydroxyacetone phosphate is the obligatory route for the synthesis of the alkyl and alkenyl lipids.

So far what has been discussed are the pathways for synthesizing phosphatidate *de novo*. This lipid can also be made by the action of diacylglycerol kinase from existing diacylglycerol. This route of metabolism is particularly important in the recycling of the diacylglycerol back into the phosphoinositides after their receptor-mediated breakdown. The possible function of phosphatidate phosphohydrolase in opposing this recycling is discussed in Chapter 2, Section V.

II. ASSAY OF PHOSPHATIDATE PHOSPHOHYDROLASE

A. Choice of Which Product to Measure

Most of the original methods for the measurement of phosphatidate phosphohydrolase activity relied upon the determination of the release of inorganic phosphate from phosphatidate emulsions. However, this method is only valid if there is no degradation of phos-

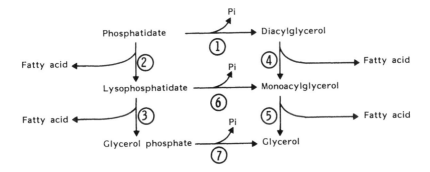

FIGURE 2. Possible routes for the degradation of phosphatidate in the phosphatidate phosphohydrolase assay. The reactions are catalyzed as follows: (1) phosphatidate phosphohydrolase; (2) phospholipase A activity; (3) lysophospholipase; (4) diacylglycerol lipase; (5) monoacylglycerol lipase; (6) probably phosphatidate phosphohydrolase; and (7) acid or alkaline phosphatase.

phatidate by phospholipases of the A type (reactions 2 and 3 of Figure 2) followed by hydrolysis of the glycerol phosphate (reaction 7). Such phospholipase A type activities have been shown to occur in the soluble and particulate fractions of rat liver (See Reference 39 and Chapter 2, Sections II and IV). The degradation of phosphatidate by soluble fractions of rat liver ends at the formation of glycerol phosphate with little further hydrolysis to glycerol and phosphate (reaction 7, Figure 2). This means that phosphatidate phosphohydrolase activity can be measured by the release of inorganic phosphate in this fraction.[43] However, the measurement of the release of water-soluble ^{32}P from diacylglycerol 3-^{32}P-phosphate would not provide a valid determination of phosphatidate phosphohydrolase activity since this could occur in glycerol phosphate. With the microsomal fraction from rat liver the glycerol phosphate is hydrolyzed by acid or alkaline phosphatases that can occur in this fraction and therefore the measurement of the production of either water-soluble ^{32}P or of inorganic phosphate would not be justified.[43] It is of course possible that inorganic phosphate could have been generated by reaction 6, Figure 2. This reaction is probably also catalyzed by phosphatidate phosphohydrolase (Section IV.C), but it requires the prior deacylation of phosphatidate to lysophosphatidate. This deacylation can be decreased in relative terms by including phosphatidylcholine with the phosphatidate in the form of a mixed micelle.[43]

The other product of the phosphohydrolase reaction is diacylglycerol which can be conveniently measured if the phosphatidate is radioactively labeled in either the glycerol or the fatty acid moieties. Labeling with glycerol also enables the rate of the deacylation reactions to be readily assayed since the radioactive glycerol phosphate or glycerol can be separated as water-soluble compounds from phosphatidate and diacylglycerol.[43] The radioactive diacylglycerol can be measured relatively simply in the chloroform phase by treating this with basic alumina.[44] Alternatively, 2 mℓ of chloroform-methanol (19:1, v/v) can be added directly to a 100-$\mu\ell$ assay system followed by 1 g of dry basic alumina. Phosphatidate, glycerol phosphate, glycerol, and fatty acid are retained with the wet alumina and the chloroform phase can be dried down so that the diacylglycerol can be measured after scintillation counting.[94]

The determination of diacylglycerol formation as a measurement of phosphatidate phosphohydrolase activity provides a valid method provided that there is little lipase activity that can degrade the diacyl- and monoacylglycerol (reactions 4 and 5 of Figure 2). This is the case in liver[43] but not in adipose tissue.[45] In the latter tissue there is no significant deacylation of the phosphatidate, but there is relatively rapid hydrolysis of diacylglycerol. Consequently, it is much better to determine phosphatidate phosphohydrolase activity in adipose tissue by following the release of inorganic phosphate. Further discussion of how to assay for phosphatidate phosphohydrolase in adipose tissue and lung is given in Chapters 3 and 5, respectively. It would be necessary in other tissues to establish whether the activity of phosphatidate phosphohydrolase should be determined by following the formation of diacylglycerol or inorganic phosphate as outlined above.

A fairly novel substrate has been used which can overcome the problems caused by the deacylase activities. This involves the use of 1-*O*-hexadecyl-*rac*-[2-^{3}H]glycerol 3-phosphate in which the ether bond is resistant to hydrolysis.[46]

B. Physical Presentation of the Substrate

Phosphatidate is an acidic phospholipid and as such the physical form in which it is presented to the phosphohydrolase is critical. This includes its salt form and the type of dispersion or membrane in which it is present. The Ca^{2+} salt form appears to be a particularly poor substrate especially for the soluble phosphatidate phosphohydrolase that requires Mg^{2+} for activity (see Section IV.A and Chapter 5, Section XIV). Ca^{2+} binds very tightly to phosphatidate and these cations are normally added to the incubation when phosphatidate is prepared from phosphatidylcholine by using phospholipase D.[47] The most effective way to remove the Ca^{2+} from phosphatidate is to treat the lipid with Chelex resin.[48] In our experience it is advantageous to acid wash the phosphatidate to remove some Ca^{2+} prior to using the Chelex resin[49] since this treatment makes the phosphatidate more soluble in chloroform-methanol-water (5:4:1 by vol) and this in turn facilitates the efficient removal of the remaining Ca^{2+}.[94] The potassium phosphatidate that is generated by this procedure gives a fairly clear dispersion in water after sonicating for a few seconds. By contrast, Ca^{2+} phosphatidate is difficult to disperse and it gives cloudy suspensions after sonication.

As explained in Section V, most early workers employed emulsions of phosphatidate that probably contained Ca^{2+} and they failed to detect phosphatidate phosphohydrolase in the cytosolic fraction of cells. This enzyme was finally discovered by using phosphatidate that was produced biosynthetically on natural membranes in the presence of Mg^{2+}. It therefore appeared that the phosphatidate needed to be presented in its natural membrane environment before it could be hydrolyzed by the soluble phosphohydrolase. However, there were in addition other phosphohydrolase activities in particulate fractions that could be detected with the phosphatidate emulsions. Since the soluble phosphohydrolase appeared to be required for the conversion of the membrane-bound phosphatidate to diacylglycerol and the particulate enzymes were relatively inactive in this respect, it became customary for several years to determine the phosphohydrolase activity by using membrane-bound substrate.

There were several disadvantages to this procedure. First, it was difficult to produce phosphatidate on the membranes without also making a small proportion of diacyglycerol since it was difficult to inhibit entirely the phosphohydrolase activity. The diacylglycerol thus produced a high blank in subsequent incubations. Secondly, the substrate was relatively undefined in terms of lipid composition and phosphatidate concentration. It was difficult to perform standard assays under very strictly controlled conditions including a known content of divalent cations. It is now known that phosphohydrolase activity can be efficiently detected in cytosolic fractions provided that the majority of the Ca^{2+} is removed from phosphatidate preparations.[1] Furthermore, this form of phosphatidate can then be incorporated into a mixed micelle with phosphatidylcholine to produce a substrate that more closely resembles the natural one. The inclusion of the phosphatidylcholine results in a stimulation of the phosphohydrolase activity.[43,50] This form of phosphatidate provides a well-defined substrate and enables phosphatidate phosphohydrolase activity to be determined under well defined conditions (see also Chapter 5, Section XV).

III. PURIFICATION OF PHOSPHATIDATE PHOSPHOHYDROLASE

Further progress towards understanding the mechanisms that control the activity will depend upon an ability to purify this enzyme or enzymes and to raise antibodies. So far this has proven to be difficult from mammalian sources.

Sedgwick and Hübscher[51] solubilized a phosphatidate phosphohydrolase from rat liver mitochondria by repeatedly freezing and thawing these organelles. The enzyme was then purified by ammonium sulfate fractionation, gel filtration, chromatography on DEAE cellulose, and finally by ultrafiltration. The increase in specific activity was 16-fold and the phosphohydrolase was stimulated by about 1.5-fold by Mg^{2+}.

Specific and nonspecific phosphatidate phosphohydrolase activities have been purified from a lyophilized microsomal fraction of rat liver.[52] The method involves lipid depletion, extraction with deoxycholate, ammonium sulfate fractionation, and chromatography on DEAE cellulose. The increase in specific activity of the specific phosphatidate phosphohydrolase was about sixfold. This enzyme preparation did not exhibit a requirement for Mg^{2+}.

Hosaka et al.[50] achieved a 15- to 20-fold purification of the soluble phosphatidate phosphohydrolase. Their method was based upon adsorption and elution from calcium phosphate, ammonium sulfate precipitation, and gel filtration. The partially purified enzyme was almost entirely dependent upon Mg^{2+} for its activity.

Later attempts to purify the cytosolic phosphatidate phosphohydrolase, which is stimulated by Mg^{2+} involved feeding rats with ethanol 6 to 8 hr before taking the livers.[53,54] This procedure resulted in an increase of more than fivefold in the phosphohydrolase activity (Chapter 2, Section II.B) which is thought to be mediated through a synthesis of new enzyme protein (Chapter 2, Section III). The initial procedure involved purifications by chromatography on DEAE cellulose, ammonium sulfate precipitation, chromatography on hydroxyapatite, a second ammonium sulfate precipitate, and finally chromatography on octyl-Sepharose. The final purification was about 50-fold. It was thought at this time that an antibody had been successfully prepared against the phosphohydrolase. However, it was then quickly realized that other preparations of control sera from different sheep also inhibited the phosphohydrolase activity. Furthermore, a strict stoichiometry did not exist between the amount of antibody added and the degree of inhibition observed with the phosphohydrolase. Consequently a correction was published.[53]

Further methods were developed for purifying the soluble phosphohydrolase so that the highest increase in specific activity that was obtained was about 416-fold.[54] The procedure involved absorption on calcium phosphate gel followed by chromatography on DEAE-cellulose, Ultrogel AcA-34, and CM-Sepharose 4B. The purification method specifically avoided precipitation of the phosphohydrolase for the fear that it might irreversibly aggregate. Tween 20 was also added to stabilize the phosphohydrolase with the hope that this would also prevent aggregation. Phosphate buffer was also found to stabilize and stimulate the activity.[54]

The molecular weight distributions of phosphatidate phosphohydrolase from the cytosolic fraction of rat liver or from microsomal fractions that had been sonicated or treated with Tween 20 have been characterized.[55] The method employed ammonium sulfate precipitation followed by chromatography on Bio-Gel A-5 m. The activity was detected by using microsomal membranes containing phosphatidate, phosphatidate dispersed by sonication with microsomal lipids, or in solvent-disrupted microsomes. The results demonstrate that the molecular weight profile of the phosphohydrolase activity is very dependent on the physical state of the substrate (see also Section II.B) and that the phosphohydrolase activity may form into aggregates. Similar conclusions are described for the phosphohydrolase activity in lung. The elution profile of phosphohydrolase activity from gel filtration columns can also be modified by incubating the enzyme preparation with phospholipids.[54]

The hydrophobic nature of the soluble phosphatidate phosphohydrolase activity from adipose tissue has been employed to characterize and purify the enzyme. Chromatography on butyl-agarose yielded four phosphohydrolase-containing fractions.[55] These components had the same pH optima and requirement for Mg^{2+} and they were inhibited by N-ethylmaleimide. However, they differed from each other in terms of hydrophobicity, sedimentation behavior, Stokes diameters, thermolability, and susceptibility to protease treatment. The four components all contained phospholipids and enzyme activity was lost when these lipids were removed.[56] It is not entirely clear at this stage of the work whether there are different peptides that are responsible for the four fractions of phosphatidate phosphohydrolase activity or whether the differences in properties can be explained by the state of aggregation of the enzyme and its association with lipid.

Details of the purification of phosphatidate phosphohydrolase activities from *Saccharomyces cerevisiae* and from lung are given in Chapter 4, Section VIII and Chapter 5, Section XIV, respectively.

IV. PROPERTIES OF PHOSPHATIDATE PHOSPHOHYDROLASE

A. Effects of Inorganic Cations and Polyamines

One of the features that has been used to characterize different phosphatidate phosphohydrolases is whether or not they are stimulated by Mg^{2+}. Generally, the cytosolic phosphatidate phosphohydrolase activities show a high dependency[49,50,57,58] for Mg^{2+} whereas those in particulate fractions are stimulated to a much smaller extent.[57,58] The whole question of the requirement of different phosphatidate phosphohydrolases is complicated because of the ability of the phosphatidate to bind divalent cations very tightly. Wherever possible the phosphatidate should be treated with Chelex resin to convert it to the K^+ salt form.[48] This substrate can then be used to determine whether indeed the phosphohydrolase is stimulated by divalent inorganic cations. However, it should also be remembered that subcellular fractions themselves contain divalent cations; for example Ca^{2+} is concentrated in the endoplasmic reticulum. Furthermore, the homogenizing medium can also be contaminated as in the case of sucrose which may contain Ca^{2+}. It is therefore advisable to treat all suspending media and other reagents where possible with cation exchange resins.

This procedure was adopted in the experiment shown in Figure 3 and the phosphatidate was treated with a Chelex type resin. The substrate was prepared as a mixed micelle at a concentration of 3 mM phosphatidate and 2 mM phosphatidylcholine. Sonication was performed in 5 mM EDTA and 5 mM EGTA. This substrate was then diluted fivefold in the assay. The high concentrations of EDTA and EGTA were chosen so as to remove any Mg^{2+} and Ca^{2+} remaining in the substrate, in other reagents, or that were present in the enzyme preparations.[94] Figure 3 shows that there was relatively little phosphohydrolase activity in either of the microsomal or cytosolic preparations in the absence of Mg^{2+}

The other purpose of this experiment was to investigate the Mg^{2+} requirement of the cytosolic phosphohydrolase when it translocates to the endoplasmic reticulum as a result of incubating these membranes with oleic acid (Chapter 2, Section IV.B). In order to do this the microsomal fraction was washed with 10 mg albumin per milliliter and then recovered by a second centrifugation step.[94] This has the effect of removing endogenous fatty acids from the membranes and displacing phosphatidate phosphohydrolase[59] as can be seen in Figure 3a. The cytosolic fraction was centrifuged twice to effect a more complete removal of particulate material. The translocation of the cytosolic phosphohydrolase was performed by mixing the washed microsomal fraction with the cytosolic fraction and incubating for 10 min at 37°C with 0.75 mM oleate. The second cytosol and the microsomes were then separated by centrifugation. Incubating with oleate resulted in a decrease in the cytosolic activity of the phosphohydrolase which was paralleled by the appearance of this enzyme in the micro-

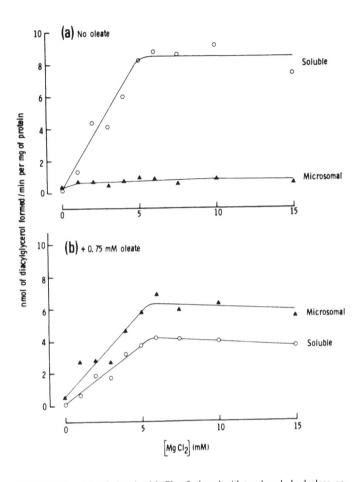

FIGURE 3. Stimulation by MgCl₂ of phosphatidate phosphohydrolase activity in microsomal and cytosolic fractions from rat liver before and after an oleate-induced translocation. A microsomal and soluble fraction was prepared from rat liver and the cytosolic reaction was recentrifuged to minimize its contamination by membranes. The microsomal fraction was resuspended in 0.25 *M* sucrose containing 0.2 m*M* dithiothreitol and 10 mg of fatty acid poor bovine serum albumin per milliliter, incubated for 10 min at 37°C and then reisolated by centrifugation. This was to remove fatty acid from the microsomal membranes and to displace phosphatidate phosphohydrolase activity from them.[59] The microsomal and soluble fractions were then incubated separately for 10 min at 37°C in 0.25 *M* sucrose containing 20 m*M* HEPES, pH 7.4, and 0.2 m*M* dithiothreitol (a). Alternatively, the microsomal and soluble fractions were recombined and incubated in the same buffer which this time contained 0.75 m*M* oleate. After cooling to 4°C the soluble and microsomal fractions were isolated again by centrifugation. The Mg²⁺ dependency of the microsomal (▲) and soluble fractions (○) was then determined and expressed relative to the protein concentration of the fractions. The total recovery of phosphohydrolase activity in (b) compared with (a) was about 70%. (From similar work by Martin, A., Hales, P., and Brindley, D. N.[94])

somal fraction (Figure 3). The phosphohydrolase in the microsomal and soluble fractions exhibited a similar marked stimulation by Mg²⁺. These results indicate that the cytosolic phosphohydrolase does not lose its Mg²⁺ requirement when it becomes bound to membranes and further confirm that the enzyme translocates.

The stimulation of the phosphohydrolase by Mg²⁺ in this and other work could occur for at least three reasons. First, the phosphohydrolase specifically recognizes the Mg²⁺ chelate of phosphatidate as a substrate. This possibility is discussed further in Section IV.C. Secondly, divalent cations have a profound effect on the packing arrangement and spacing of phosphatidate molecules in membranes.[60] The correct concentration of Mg^{2+} could thus create the appropriate physical conditions in the membranes such that the phosphohydrolase is able to interact, insert itself into the membrane, and then hydrolyze the phosphatidate. The effect of Mg^{2+} on the organization of phosphatidate in membranes is known to be different[61,62] from that of Ca^{2+}. Finally, a requirement for Mg^{2+} could occur in order to compete with Ca^{2+} attached to the phosphatidate and to remove its inhibitory effects. This possibility is a less likely explanation for the results in Figure 3 since great care was taken to exclude a Ca^{2+} effect, but it could be a contributing factor in other work.

Co^{2+} can partially substitute for Mg^{2+} in stimulating the activity of phosphatidate phosphohydrolase but Mn^{2+}, Fe^{2+}, and Fe^{3+} were much less effective.[49,63,64] Ca^{2+}, Zn^{2+}, Co^{2+}, Ni^{2+}, and Mn^{2+} also antagonize[45,57,58] the stimulatory effect of Mg^{2+}. Experiments have also been conducted[53] to see whether the loss of phosphohydrolase activity that is seen on purification of the enzyme can be regained by the addition of Mg^{2+}, Mn^{2+}, or Zn^{2+}, but the results were negative. The phosphohydrolase was also treated with 1,10-phenanthroline to see whether its activity depended upon tightly bound Zn^{2+} as is the case with other phospholipases of the C type. However, the results were again negative.[54]

Polyamines had relatively little effect[49] at stimulating the phosphohydrolase activity in the absence of Mg^{2+}, but they may potentiate the effects of Mg^{2+} (see also Chapter 3, Section III.D.8). A further effect on the phosphohydrolase activity could occur at the level of the translocation of the soluble phosphohydrolase to the endoplasmic reticulum since polyamines appear to facilitate the action of fatty acids in this respect (Chapter 2, Section IV.C).

B. Effects of Amphiphilic Ions on Phosphatidate Phosphohydrolase Activity

A variety of drugs including some phenothiazine neuroleptics, imipramine, antidepressants, local anesthetics, anorectics, hypolipidemic agents, some β-blockers, morphine, and levorphenol can inhibit the synthesis of triacylglycerol and phosphatidylcholine (Chapter 2, Section III.H) by blocking the action of phosphatidate phosphohydrolase (see Chapter 2, Table 3). Despite the diversity of pharmacological function, these compounds all possess a hydrophobic region and a primary or substituted amine. These features enable the drugs to interact with lipid membranes and the negative charge on the phosphate groups of phosphatidate respectively (see Chapter 2, Sections II.H and IV.D). The interaction of the drugs with the membrane changes its physical characteristics and thus the ability of phosphatidate phosphohydrolase to act on its substrate.

Experiments with phosphatidate emulsions showed that the addition of an amphiphilic amine, chlorpromazine, in the absence of Mg^{2+} stimulated the soluble phosphatidate phosphohydrolase activity (Figure 4). The extent of the stimulation was similar to that observed in the presence of optimum concentrations of Mg^{2+}. If Mg^{2+} was added then the stimulating effect of the chlorpromazine was decreased (Figure 4) and higher concentrations of chlorpromazine inhibited the reaction.[49,58] These results therefore showed an interdependence of the effects of chlorpromazine and Mg^{2+} which modify the interaction of the phosphohydrolase with its substrate.[49] For example, the drug could change the charge on the surface of the phosphatidate emulsion from negative to positive, or it could alter the packing arrangement of the membrane.[60] Further experiments[49] also demonstrated that chlorpromazine could stimulate the phosphohydrolase activity in the presence of up to 24 mM EDTA and in the absence of added Mg^{2+}. These results therefore imply that Mg^{2+} may not be an absolute requirement for the phosphohydrolase, but that what is needed is a substrate that is presented to the enzyme in a suitable form.[49] A further effect of the chlorpromazine could have been to displace traces of Ca^{2+} from the phosphatidate[60] but the substrate in these experiments[49] was treated with Chelex resin.

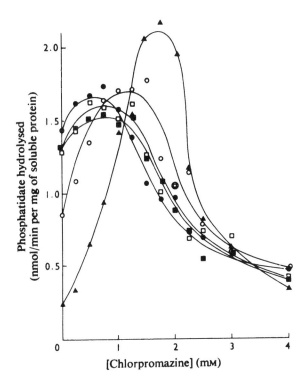

FIGURE 4.   Effects of Mg^{2+} and chlorpromazine on the activity of phosphatidate phosphohydrolase. The concentrations of Mg^{2+} in the assays were ▲, 0 mM; ○, 0.3 mM; ■, 0.6 mM; □, 1 mM and ●, 1.5 mM. (From Bowley, M. et al., *Biochem. J.*, 165, 447, 1977. With permission.)

In subsequent work the phosphatidate emulsion was replaced by a mixed emulsion of phosphatidylcholine with the phosphatidate. The addition of phosphatidylcholine in itself stimulated the phosphohydrolase activity (Section II.A). In this situation chlorpromazine had relatively little stimulating effect in the presence of EDTA and EGTA[94] compared to the large stimulation that was produced by Mg^{2+} (Figure 3). However, in the absence of phosphatidylcholine, chlorpromazine was able to produce a stimulation as was observed before (Figure 4). These results imply first that the production of a more natural membrane by the addition of phosphatidylcholine masked the stimulating effect of chlorpromazine in the absence of Mg^{2+}. Secondly, the Mg^{2+} dependency of the phosphohydrolase is very marked when the phosphatidylcholine-phosphatidate emulsion is used and this dependency cannot be replaced by chlorpromazine. These results provide further evidence that the use of aqueous dispersions of phosphatidate may be measuring a different phosphohydrolase activity than that detected when the phosphatidate is incorporated into a more natural membrane (see Reference 55 and Chapter 5, Section XIV).

Amphiphilic anions such as clofenapate or oleoyl-CoA can have the opposite effect of increasing phosphatidate phosphohydrolase activity if there is excess Mg^{2+} present in the substrate (see Chapter 2, Figure 16 and Section V). In other work, fatty acids and their acyl-CoA esters have been found to inhibit the Mg^{2+}-dependent phosphohydrolase activity (Reference 51 and Chapter 3, Section III.D.10). However, this action may not be physiologically relevant. In fact, it appears more likely that fatty acids should stimulate the expression of phosphatidate phosphohydrolase activity in vivo by facilitating the attachment of the soluble enzyme to the endoplasmic reticulum (Chapter 2, Section IV.B, Chapter 3, Section VI, and Chapter 5, Section XVI).

C. Substrate Specificity

Phosphatidate phosphohydrolase from hamster small intestine,[65] rat liver,[52] and the Mg^{2+} insensitive activity in lung (Chapter 5, Section XIII) has been reported to hydrolyze lysophosphatidate in addition to phosphatidate. It is, however, difficult to be absolutely certain about this specificity since relatively crude preparations of the enzyme were employed and other contaminating activities could be responsible for the hydrolysis of the lysophosphatidate. Hosaka et al.[50] used preparations of Mg^{2+}-stimulated phosphohydrolase that were purified from the soluble fraction of rat liver and found that the rate of hydrolysis of lysophosphatidate was about 10% that of phosphatidate.

The phosphohydrolase from brain can hydrolyze monoalkylphosphate in addition to phosphatidate but it was reported not to hydrolyze dialkylphosphates or synthetic dipalmitoylphosphatidate.[66] The reason for the lack of effect on the latter compound is unexpected and it may have resulted from a problem in the physical presentation of the substrate. Hexadecylphosphate was also found to be a substrate for a partially purified phosphatidate phosphohydrolase from rat liver[51] and 1-*O*-hexadecyl-*rac*-[2-³H]glycerol 3-phosphate is hydrolyzed by human amniotic fluid.[46] A preparation from erythrocytes was able to hydrolyze 1,3-diacylglycerol 2-phosphate.[67] It has also been suggested that phosphatidylglycerol phosphate is a substrate for phosphatidate phosphohydrolase and this will be discussed in greater detail in Chapter 5, Sections XIII and XVII.

Other phospholipids such as phosphatidylcholine,[51,65] phosphatidylethanolamine,[65] diphosphatidylglycerol,[66] phosphatidylinositol 4-phosphate,[64] and phosphatidylinositol 4,5-bisphosphate[51,64,68] are not substrates for the phosphohydrolase. Water-soluble phosphate esters are also not substrates.[50-52,58]

There is evidence that the fatty acid composition of the phosphatidate may also influence the activity of the phosphohydrolase.[69] This work employed phosphatidate phosphohydrolase preparations from brain and showed that the hydrolysis of synthetic phosphatidate containing two oleic acid groups was about 50% lower than were the rates with phosphatidate prepared from phosphatidylcholine of soybean, yeast, or eggs.

Indirect evidence for the preferential hydrolysis of that phosphatidate containing unsaturated fatty acids in addition to palmitate was obtained from work with rat liver mitochondria and the microsomal fraction from the small intestinal mucosa of cats.[70] Phosphatidate was generated from glycerol phosphate and different combinations of fatty acids and it was found that the unsaturated acids promoted the conversion of phosphatidate to diacylglycerol. This work was pursued by labeling the membranes from these tissues with phosphatidates of different fatty acid compositions.[71] Phosphatidate containing myristic and palmitic acids was a better substrate for the soluble phosphohydrolase than that containing lauric and stearic acids. Furthermore, the phosphatidate that was synthesized from a mixture of palmitic and oleic acids was hydrolyzed at a higher rate than phosphatidate containing these acids separately. By contrast, the production of diacylglycerol by the residual phosphohydrolase that was present in the membranes in the absence of the soluble fraction was highest when the phosphatidate was prepared from stearate, oleate, or a mixture of palmitate and oleate.[71] Such a preference for phosphatidate with a saturated fatty acid presumably at the 1-position and an unsaturated fatty acid at the 2-position could help cells to produce phosphatidylcholine and phosphatidylethanolamine with their expected fatty acid compositions. However, doubt has been cast on whether this specificity of phosphatidate phosphohydrolase is really expressed in the liver in vivo.[72,73]

Many investigators have presented information concerning the apparent K_m values for phosphatidate phosphohydrolase from various sources. Generally, these values lie in the range 10 μM to 1 mM: the higher values normally being found with phosphatidate emulsions and lower values associated with phosphatidate incorporated into membranes or mixed with other phospholipids (References 44, 49 to 52, 54, and 58, Chapter 3, Section III.D.1,

Chapter 4, Section III, and Chapter 5, Table 3). However, it is difficult to interpret exactly what these values mean since the physical state of the substrate and its content of metal ions has such a profound effect on K_m. For example, phosphatidate could be in micelles, unilamellar vesicles, or in membrane sheets. The K_m value only reflects the concentration of those molecules of phosphatidate that are accessible to the phosphohydrolase. Changes in the apprent K_m value have been observed at different stages of enzyme purification,[54] but this could also depend upon the state of aggregation of the phosphohydrolase or the presence of detergent in the enzyme preparation.

A nonhydrolyzable phosphatidate analog, 2-hexadecoxy-3-octadecoxypropylphonate was found to be an inhibitor of phosphatidate phosphohydrolase from pig kidney. At equimolar concentration with phosphatidate it was able to inhibit the reaction by 96%. In order to obtain a maximum inhibition it was necessary to preincubate the enzyme with the inhibitor.

D. pH Optima

The effect of pH on the activity of various phosphatidate phosphohydrolase activities is discussed in Chapter 3, Section III, Chapter 4, Sections III and V, and Chapter 5, Sections XIII and XIV. In liver a pH optimum for phosphohydrolase activity has been reported to occur at about 6 when mitochondrial fractions were used.[75] However, the Mg^{2+}-stimulated activity that is primarily isolated in the soluble fraction had a fairly broad pH optimum[50,75] between 6 to 7.5. The two phosphohydrolase fractions that were isolated from microsomal fractions of rat liver had broad pH profiles with optimas[52] in the range 6.7 to 6.8. These activities, however, were not stimulated by Mg^{2+}

E. Effects of Fluoride and Reagents that React with Sulfhydryl and Arginine Groups

It is well known that F^- inhibits the activities of phosphatidate phosphohydrolases.[1,76] This property has been exploited during the preparation of membranes labeled biosynthetically with phosphatidate. The addition of 50 mM F^- largely prevents the subsequent conversion of the phosphatidate to diacylglycerol.

Sulfhydryl groups appear to be required for the activity of the Mg^{2+}-dependent phosphatidate phosphohydrolase. This activity can be inhibited completely by N-ethylmaleimide[54] and partially by p-chloromercuribenzoate.[54] Iodoacetate was ineffective under the conditions employed in this work.[54] Furthermore, phosphohydrolase activity from rat lung was also not inhibited by iodoacetamide although p-chloromercuriphenylsulfonate was effective.[77] Similarly, the microsomal phosphohydrolase activity of rat liver was inhibited by p-chloromercuribenzoate but not by iodoacetate.[68] Hosaka et al.[50] also found that the Mg^{2+}-stimulated phosphohydrolase activity could be inhibited by 90% in the presence of 50 μM p-chloromercuribenzoate and the absence of dithiothreitol. The lower effect of p-chloromercuribenzoate observed by Butterwith et al.[54] is probably explained by the presence of 0.2 mM dithiothreitol in their assays, although an excess of the thiol-blocking reagents was used. It is possible that part of the effect of p-chloromercuribenzoate could also have involved the binding of the mercury to a metal binding site on the enzyme or its substrate. $HgCl_2$ is known to inhibit the phosphohydrolase activity from lung.[78,79]

The Mg^{2+}-independent phosphatidate phosphohydrolase activity of rat adipose tissue is resistant to inhibition by thiol-blocking reagents. It has therefore been proposed that the use of N-ethylmaleimide may be a convenient method for distinguishing between the Mg^{2+}-sensitive and Mg^{2+}-insensitive phosphohydrolase activities (Chapter 3, Section III.D.5).

The inhibition by butane-2,3-dione and cyclohexane-1,2-dione of the Mg^{2+}-stimulated phosphatidate phosphohydrolase activity, that had been partially purified from rat liver indicates that this enzyme also requires arginine residues for catalytic activity. This group is also required by several other enzymes that act upon anionic substrates, and a similar effect has been shown with the phospholipase C activity of *Bacillus cerius*.[80]

F. Effects of Nonionic Detergents

The effects of anionic and cationic detergents on phosphatidate phosphohydrolase activity have already been discussed in Section IV.B. Nonionic detergents also modify the phosphohydrolase activity in ways which depend very much on the assay conditions that are used. Ide and Nakazawa[55] demonstrated that Tween 20 increased the activity of a cytosolic phosphohydrolase by about twofold when it was assayed with either membrane-bound phosphatidate or phosphatidate in sonicated dispersions of solvent disrupted microsomes. By contrast Tween 20 inhibited the activity towards membrane-bound phosphatidate when this was presented in the presence of an optimum concentration of Mg^{2+}. If the phosphohydrolase activity was measured with an aqueous dispersion of phosphatidate, Tween 20 had little effect on the enzyme activity.[55] Jamdar et al.[56] have also demonstrated that both Tween 20 and Triton X-100 can inhibit the various forms of the Mg^{2+}-stimulated phosphohydrolase that they isolated from the cytosol of rat adipose tissue.

Further information concerning the effect of Triton X-100 on the activity of the Mg^{2+}-dependent and Mg^{2+}-independent phosphatidate phosphohydrolases can be obtained from Chapter 5, Sections XIV and XV. It may be possible to use the differential effects of detergents to distinguish between these activities provided that the physical state of the substrate that is used is very carefully controlled.

Apart from these effects of neutral detergents on the measurement of phosphatidate phosphohydrolase activity, it is also known that Tween 20 can stabilize the activity[54] and alter its apparent molecular size as judged by gel filtration.[55]

V. SUBCELLULAR DISTRIBUTION OF PHOSPHATIDATE PHOSPHOHYDROLASE ACTIVITY

Phosphatidate phosphohydrolase activity has been reported to occur in plasma membranes,[81] mitochondria, lysosomes, the microsomal fraction,[82,83] and the soluble fraction of the liver.[1,76] In two of these studies the highest specific activity was, in fact, recorded in the lysosomal fraction[82,83] but presumably this activity would be involved in the degradation rather than in the synthesis of glycerolipids.

Solubilization of the phosphohydrolase from the microsomal fraction of rat liver has shown that it can be resolved into two distinct activities.[52] One fraction (F_A) was fairly nonspecific in that it hydrolyzed several phosphate esters. It has a relatively high K_m for phosphatidate and it was not inhibited by diacylglycerol. Fraction F_B was specific for the hydrolysis of phosphatidate or lysophosphatidate. It had a relatively low apparent K_m for phosphatidate and it was inhibited in a noncompetitive manner by diacylglycerol. It was thought that this enzyme was the one involved in glycerolipid synthesis.[52]

The phosphohydrolase in mitochondria can be separated into activity that can be readily extracted by repeatedly freezing and thawing and activity that remained membrane bound.[51] Further purification of the solubilized activity gave a preparation that could hydrolyze hexadecylphosphate, glycerol 2-phosphate, ATP, and phosphatidate. However, evidence was presented that hexadecylphosphate and phosphatidate were hydrolyzed by a different enzyme from that which acted upon the water-soluble substrates.

In early work on the subcellular distribution of phosphatidate phosphohydrolase many workers failed to detect activity in the soluble fraction. This was probably caused by the use of emulsions that contained phosphatidate in its Ca^{2+} salt form (Section II). It is extremely difficult to remove Ca^{2+} since it binds very tightly to phosphatidate. The Ca^{2+} inhibits the soluble phosphohydrolase activity which requires the phosphatidate to be in the Mg^{2+}-salt form. The soluble phosphohydrolase was originally described as a factor that was present in the particle-free supernatant of tissues and which stimulated the synthesis of diacylglycerol and triacylglycerol by particulate fractions.[84-87] For many years its nature remained a mystery.

This was because conventional wisdom dictated that all of the phosphatidate phosphohydrolase activity was tightly membrane bound. Its discovery relied on synthesis of phosphatidate on membranes from glycerol phosphate and the observation that the soluble fraction then converted this to diacylglycerol. This activity was destroyed by heating the soluble fraction.[75,88] The phosphatidate in the membranes was, of course, generated in the presence of Mg^{2+} rather than Ca^{2+}. Following this it became customary to determine the activity of the soluble enzyme by using natural membranes that contained biosynthetically produced phosphatidate.[1,76] However, it is now known that the activity of soluble phosphohydrolase can be efficiently determined by using an artificial membrane composed of phosphatidylcholine and phosphatidate (see Section II).

The physiological function of the cytosolic phosphohydrolase has been the subject of much debate since the other enzymes that are directly involved in glycerolipid synthesis are tightly membrane bound. At some time this soluble phosphohydrolase must come into contact with these membranes in order to interact with the phosphatidate. This should present little problem in the case of phosphatidate produced in the endoplasmic reticulum since the synthesis takes place on the cytoplasmic surface (Section I). The situation is less clear for mitochondrially produced phosphatidate because it is not entirely certain whether the synthesis occurs on the outer or inner surface of the outer mitochondrial membrane. It is also possible that phosphatidate phosphohydrolase is only loosely attached to particulate fractions and it becomes detached during homogenization and centrifugation. Evidence in support of this suggestion is provided by the observation that the proportion of membrane-bound phosphatidate phosphohydrolase can be changed by altering the ionic strength of the homogenizing medium.[89] Alternatively, it is possible that the ability of cells to control the reversible binding of the phosphohydrolase to the membranes where phosphatidate is synthesized could help to control the rate and direction of glycerolipid synthesis.[90] This type of movement of a regulatory enzyme between cell compartments is typical of an ambiquitous enzyme.[91] Evidence that phosphatidate phosphohydrolase is an ambiquitous enzyme was first provided by Moller and his colleagues[92] in adipose tissue by studying the effects of a series of lipolytic hormones (see Chapter 3, Section VI.B). It has subsequently been shown that fatty acids and their CoA esters are probably the most important factors in controlling the interaction of the soluble phosphohydrolase with the endoplasmic reticulum[93] and that this effect is enhanced, at least in vitro, by polyamines (Chapter 2, Sections IV.C and VI.A).

The implications of these results are that the dilution of cell homogenates during fractionation could lead to the dissociation of the phosphohydrolase from membranes. Ideally, subcellular fractionation should be performed on homogenates treated with different concentrations of fatty acids in order to assess which of the possible multiple forms phosphatidate phosphohydrolase can interact within the various membranes of the cell. One method that has been used fairly widely to distinguish between various phosphatidate phosphohydrolases is their dependency on Mg^{2+}. Generally, the soluble phosphohydrolase exhibits a much higher requirement for Mg^{2+} than phosphatidate phosphohydrolase activities in the microsomal or mitochondrial fractions (Section IV.A). This distinction of Mg^{2+}-dependent and Mg^{2+}-independent activities in plant tissues, lung, and adipose tissue is dealt with in Chapters 2, 4, and 5 respectively. There are several problems in understanding clearly the reported Mg^{2+} requirements of the phosphohydrolase present in various fractions from the liver. First it is not always clear whether the Ca^{2+} and Mg^{2+} contents of the phosphatidate substrate were absolutely controlled (see also Sections II and IV). Secondly, the liver contained phospholipases of the A type that degrade phosphatidate. Consequently, if the phosphohydrolase activity was determined by measuring the release of inorganic phosphate, then the assay can be invalid (Section II). It therefore appears that more work needs to be done in liver to clearly define the subcellular location of phosphatidate phosphohydrolase activity and the extent to which true Mg^{2+}-independent activity is present in particular cells fractions.

These studies might be aided by using thiol-group inhibitors or Triton X-100 in order to distinguish between the Mg^{2+}-sensitive and -insensitive phosphatidate phosphohydrolases (see also Chapter 3, Section III.D.5 and Chapter 5, Sections XIV and XV).

The work shown in Figure 3 demonstrates that the Mg^{2+}-independent activity in washed microsomal fractions is relatively small and that an Mg^{2+}-dependent phosphohydrolase from the cytosol can be made to attach to the membranes when they are incubated with oleate.[94]

VI. CONCLUSIONS

This chapter is intended as an introduction to glycerolipid synthesis and the general properties and function of various phosphatidate phosphohydrolases. The following chapters will amplify this discussion to consider the specific requirements for glycerolipid synthesis in plants and microorganisms, and in various tissues primarily from mammals.

REFERENCES

1. **Brindley, D. N. and Sturton, R. G.,** Phosphatidate metabolism and its relation to triacylglycerol biosynthesis, in *Phospholipids. New Comp. Biochem.* 4, 179, 1982.
2. **Coleman, R. and Bell, R. M.,** Evidence that the biosynthesis of phosphatidylethanolamine, phosphatidylcholine and triacylglycerol occurs on the cytoplasmic side of microsomal vesicles, *J. Cell Biol.,* 76, 245, 1978.
3. **Hajra, A. K.,** Biosynthesis of acyldihydroxyacetone phosphate in guinea pig liver mitochondria, *J. Biol. Chem.,* 243, 3458, 1968.
4. **Schlossman, D. M. and Bell, R. M.,** Triacylglycerol synthesis in isolated fat cells. Evidence that the *sn*-glycerol 3-phosphate and dihydroxyacetone phosphate acyltransferases are dual catalytic functions of a single microsomal enzyme, *J. Biol. Chem.,* 251, 5738, 1976.
5. **Schlossman, D. M. and Bell, R. M.,** Microsomal sn-glycerol 3-phosphate acyltransferase activities from liver and other tissues. Evidence for a single enzyme catalysing both reactions, *Arch. Biochem. Biophys.,* 182, 732, 1977.
6. **Daae, L. N. W.,** The mitochondrial acylation of glycerol phosphate in rat liver. Fatty acid and positional specificity, *Biochim. Biophys. Acta,* 270, 23, 1972.
7. **Shephard, E. H. and Hübscher, G.,** Phosphatidate biosynthesis in mitochondrial subfractions of rat liver, *Biochem. J.,* 113, 429, 1969.
8. **Zborowski, J. and Woitczak, L.,** Phospholipid synthesis in rat liver mitochondria, *Biochim. Biophys. Acta,* 187, 73, 1969.
9. **Nimmo, H. G.,** The location of glycerol phosphate acyltransferase and fatty acyl-CoA synthetase in the inner surface of the mitochondrial outer membrane, *FEBS Lett.,* 101, 262, 1979.
10. **Hesler, C. B., Carroll, M. A., and Halder, D.,** The topography of glycerophosphate acyltransferase in the transverse plane of the mitochondrial outer membrane, *J. Biol. Chem.,* 260, 7452, 1985.
11. **Bowley, M., Manning, R., and Brindley, D. N.,** The tritium isotope effect of *sn*-glycerol 3-phosphate oxidase and the effects of clofenapate and N-(2-benzoyloxyethyl)norfenfluramine in the esterification of glycerol phosphate and dihydroxyacetone phosphate by rat liver mitochondria, *Biochem. J.,* 136, 421, 1973.
12. **Hajra, A. K., Burke, C. L., and Jones, C. L.,** Subcellular localization of acyl-CoA dihydroxyacetone phosphate acyltransferase in rat liver peroxisomes, *J. Biol. Chem.,* 254, 10896, 1979.
13. **Bates, E. J. and Saggerson, E. D.,** A study of the glycerol phosphate acyltransferase and dihydroxyacetone phosphate acyltransferase activities in rat liver mitochondrial and microsomal fractions, *Biochem. J.,* 182, 751, 1979.
14. **Stern, W. and Pullman, M. E.,** Acyl-CoA sn-glycerol 3-phosphate acyltransferase and the positional distribution of fatty acids in phospholipids of cultured cells, *J. Biol. Chem.,* 253, 8047, 1978.
15. **Yamada, K. and Okuyama, H.,** Selectivity of diacylglycerophosphate synthesis in subcellular fractions of rat liver, *Arch. Biochem. Biophys.,* 190, 409, 1978.
16. **Halder, D., Tso, W-W., and Pullman, M. E.,** The acylation of sn-glycerol 3-phosphate in mammalian organs and Ehrlich ascites tumor cells, *J. Biol. Chem.,* 254, 4502, 1979.

17. **Nimmo, H. G.**, Evidence for the existence of isoenzymes of glycerol phosphate acyltransferase, *Biochem. J.*, 177, 283, 1979.
18. **Halder, D., Carroll, M., Morris, P., Grosjean, C., and Anzalone, T.**, Further distinguishing properties of mitochondrial and microsomal glycerophosphate acyltransferase (GAT), *Fed. Proc.*, 39, 1992, 1980.
19. **Jones, C. L. and Hajra, A. K.**, Properties of guinea pig liver peroxisomal dihydroxyacetone phosphate acyltransferase, *J. Biol. Chem.*, 255, 8289, 1980.
20. **Lawson, N., Jennings, R. J., Pollard, A. D., Sturton, R. G., Ralph, S. J., Marsden, C. A., Fears, R., and Brindley, D. N.**, Effects of chronic modification of dietary fat and carbohydrate in rats. The activities of some enzymes of hepatic glycerolipid synthesis and the effects of corticotropin injection, *Biochem. J.*, 200, 265, 1981.
21. **Monroy, G., Rola, F. H., and Pullman, M. E.**, A substrate and positional-specific acylation of *sn*-glycerol 3-phosphate by rat liver mitochondria, *J. Biol. Chem.*, 247, 6884, 1972.
22. **Monroy, G., Kelker, H. C., and Pullman, M. E.**, Partial purification and properties of an acyl-CoA:*sn*-glycerol 3-phosphate acyltransferase from rat liver mitochondria, *J. Biol. Chem.*, 248, 2845, 1973.
23. **Bjerve, K. S., Daae, L. M. W., and Bremer, J.**, Phosphatidic acid biosynthesis in rat liver mitochondria and microsomal fractions, *Biochem. J.*, 158, 249, 1976.
24. **Abou-Issa, H. M. and Cleland, W. W.**, Studies on the microsomal acylation of L-glycerol-3-phosphate. II. The specificity and properties of the rat liver enzyme, *Biochim. Biophys. Acta*, 176, 692, 1969.
25. **Ray, T. K., Crona, J. E., Mavis, R. D., and Vagelos, P. R.**, The specific acylation of glycerol 3-phosphate to monoacylglycerol 3-phosphate in *Escherichia coli*, *J. Biol. Chem.*, 245, 6442, 1970.
26. **Yamashita, S., Hosaka, K., and Numa, S.**, Resolution and reconstitution of the phosphatidate synthesizing system of rat liver microsomes, *Proc. Natl. Acad. Sci. U.S.A.*, 69, 3490, 1972.
27. **Yamashita, S. and Numa, S.**, Partial purification and properties of glycerol phosphate acyltransferase from rat liver, *Eur. J. Biochem.*, 31, 565, 1972.
28. **Bremer, J., Bjerve, K. S., Borrebaek, B., and Christiansen, R.**, The glycerophosphate acyltransferases and their function in the metabolism of fatty acids, *Mol. Cell. Biochem.*, 12, 113, 1976.
29. **Ballas, L. M., Lasarow, P. B., and Bell, R. M.**, Glycerolipid synthetic capacity of rat liver peroxisomes, *Biochim. Biophys. Acta*, 795, 297, 1984.
30. **Van den Bosch, H.**, Phosphoglyceride metabolism, *Annu. Rev. Biochem.*, 43, 243, 1974.
31. **Bell, R. M. and Coleman, R. A.**, Enzymes of glycerolipid synthesis in eukaryotes, *Annu. Rev. Biochem.*, 49, 459, 1980.
32. **Molaporast-Shahidsaless, F., Shrago, E., and Elson, C. E.**, α-Glycerophosphate and dihydroxyacetone phosphate in rats fed high fat or high sucrose diets, *J. Nutr.*, 109, 1560, 1979.
33. **Pollock, R. J., Hajra, A. K., and Agranoff, B. W.**, The relative utilization of the acyl dihydroxyacetone phosphate and glycerol phosphate pathways for synthesis of glycerolipids in various tumors and normal tissues, *Biochim. Biophys. Acta*, 380, 421, 1975.
34. **Fisher, A. B., Huber, G. A., Furia, L., Bassett, D., and Rabinowitz, J. L.**, Evidence for lipid synthesis by the dihydroxyacetone phosphate pathway in rabbit lung subcellular fractions, *J. Lab. Clin. Med.*, 87, 1033, 1976.
35. **Agranoff, B. W. and Hajra, A. K.**, The acyl dihydroxyacetone phosphate pathway for glycerolipid biosynthesis in mouse liver and Ehrlich ascites tumor cells, *Proc. Natl. Acad. Sci. U.S.A.*, 68, 411, 1971.
36. **Declercq, P. E., Haagsman, H. P., Van Veldhoven, P., Debeer, L. J., Van Golde, L. M. G., and Mannaerts, G. P.**, Rat liver dihydroxyacetone phosphate acyltransferases and their contribution to glycerolipid synthesis, *J. Biol. Chem.*, 259, 9064, 1984.
37. **Patel, B. N., Mackness, M. I., and Connock, M. J.**, Effect of pyrophosphate on dihydroxyacetone phosphate acyltranferase activity in male albino mouse tissues, *Biochem. Soc. Trans.*, 13, 249, 1985.
38. **Manning, R. and Brindley, D. N.**, Tritium isotope effect in the measurement of the glycerol phosphate and dihydroxyacetone phosphate pathways of glycerolipid biosynthesis in rat liver, *Biochem. J.*, 130, 1003, 1972.
39. **Bowley, M. and Brindley, D. N.**, Selective inhibition by clofenapate of glycerolipid synthesis via the esterification of dihydroxyacetone phosphate in rat liver slices, *Int. J. Biochem.*, 7, 141, 1976.
40. **Mason, R. J.**, Importance of the acyldihydroxyacetone phosphate pathway in the synthesis of phosphatidylglycerol and phosphatidylcholine in alveolar type II cells, *J. Biol. Chem.*, 253, 3367, 1978.
41. **Pollock, R. J., Hajra, A. K., Folk, W. R., and Agranoff, B. W.**, Use of [1 or 3-^{3}H, U-^{14}C]glucose to estimate the synthesis of glycerolipids via acyldihydroxyacetone phosphate, *Biochem. Biophys. Res. Commun.*, 65, 658, 1975.
42. **Pollock, R. J., Hajra, A. K., and Agranoff, B. W.**, Incorporation of D-[3-^{3}H, U-^{14}C]glucose into glycerolipid via acyl dihydroxyacetone phosphate in untransformed and viral-transformed BHK-21-c13 fibroblasts, *J. Biol. Chem.*, 251, 5149, 1976.
43. **Sturton, R. G. and Brindley, D. N.**, Problems encountered in measuring the activity of phosphatidate phosphohydrolase, *Biochem. J.*, 171, 263, 1978.

44. **Pollard, A. D. and Brindley, D. N.**, Effect of vasopressin and corticosterone on fatty acid metabolism and on the activities of glycerol phosphate acyltransferase and phosphatidate phosphohydrolase in rat hepatocytes, *Biochem. J.*, 217, 461, 1984.

45. **Lawson, N., Pollard, A. D., Jennings, R. J., Gurr, M. I., and Brindley, D. N.**, The activities of lipoprotein lipase and of enzymes involved in triacylglycerol synthesis in rat adipose tissue, *Biochem. J.*, 2, 285, 1981.

46. **Bleasdale, J. E., Davis, C-S., and Agranoff, B. W.**, The measurement of phosphatidate phosphohydrolase in human amniotic fluid, *Biochim. Biophys. Acta*, 528, 331, 1978.

47. **Ansell, G. B. and Hawthorne, J. N.**, Preparation of phospholipids, *B.B.A. Libr.*, 3, 85, 1964.

48. **Renkonen, O.**, Mono- and dimethylphosphatides from different subtypes of choline and ethanolamine glycerophosphatides, *Biochem. Biophys. Acta*, 152, 114, 1969.

49. **Bowley, M., Cooling, J., Burditt, S. L., and Brindely, D. N.**, The effects of amphiphilic cationic drugs and inorganic cations on the activity of phosphatidate phosphohydrolase, *Biochem. J.*, 165, 447, 1977.

50. **Hosaka, K., Yamashita, S., and Numa, S.**, Partial purification, properties and subcellular distribution of rat liver phosphatidate phosphohydrolase, *J. Biochem.*, 77, 501, 1975.

51. **Sedgwick, B. and Hübscher, G.**, Metabolism of phospholipids. X. Partial purification and properties of a soluble phosphatidate phosphohydrolase from rat liver, *Biochem. Biophys. Acta*, 144, 397, 1967.

52. **Caras, I. and Shapiro, B.**, Partial purification and properties of microsomal phosphatidate phosphohydrolase from rat liver, *Biochem. Biophys. Acta*, 409, 201, 1975.

53. **Sturton, R. G., Butterwith, S. C., Burditt, S. L., and Brindley, D. N.**, Effects of starvation, corticotropin injection, and ethanol feeding on the activity and amount of phosphatidate phosphohydrolase in rat liver, *FEBS Lett.*, 126, 297, 1981; correction added in *FEBS Lett.*, 130, 317, 1981.

54. **Butterwith, S. C., Hopewell, R., and Brindley, D. N.**, Partial purification and characterisation of the soluble phosphatidate phosphohydrolase of rat liver, *Biochem. J.*, 220, 825, 1984.

55. **Ide, H. and Nakazawa, Y.**, Phosphatidate phosphatase in rat liver: the relationship between the activities with membrane bound phosphatidate and aqueous dispersion of phosphatidate, *J. Biochem.*, 97, 45, 1985.

56. **Wells, G. N., Osborne, L. J., and Jamdar, S. C.**, Isolation and characterisation of multiple forms of Mg^{2+}-dependent phosphatidate phosphohydrolase from rat adipose cytosol, *Biochim. Biophys. Acta*, 878, 225, 1986.

57. **Jamdar, S. C. and Fallon, H. J.**, Glycerolipid synthesis in rat adipose tissue. II. Properties and distribution of phosphatidate phosphatase, *J. Lipid Res.*, 14, 517, 1973.

58. **Sturton, R. G. and Brindley, D. N.**, Factors controlling the metabolism of phosphatidate by phosphohydrolase and phospholipase A-type activities, *Biochim. Biophys. Acta*, 619, 494, 1980.

59. **Hopewell, R., Martin-Sanz, P., Martin, A., Saxton, J., and Brindley, D. N.**, Regulation of the translocation of phosphatidate phosphohydrolase between the cytosol and the endoplasmic reticulum of rat liver, *Biochem. J.*, 232, 485, 1985.

60. **Cullis, P. R. and DeKruijff, B.**, Lipid polymorphism and the functional roles of lipids in biological membranes, *Biochim. Biophys. Acta*, 559, 399, 1979.

61. **Ito, T. and Ohnishi, S.**, Ca^{2+}-induced lateral phase separations in phosphatidic acid-phosphatidylcholine membranes, *Biochim. Biophys. Acta*, 352, 29, 1974.

62. **Miner, V. W. and Prestegard, J. H.**, Structure of divalent cation-phosphatidic acid complexes as determined by ^{31}P-NMR, *Biochim. Biophys. Acta*, 774, 227, 1984.

63. **Roncari, D. A. K. and Hollenberg, C. H.**, Esterification of free fatty acids by subcellular preparations of rat adipose tissue, *Biochim. Biophys. Acta*, 137, 446, 1967.

64. **Hosaka, K. and Yamashita, S.**, Partial purification and properties of phosphatidate phosphatase in *Saccharomyces cerevisiae*, *Biochim. Biophys. Acta*, 796, 102, 1984.

65. **Johnston, J. M. and Bearden, J. H.**, Intestinal phosphatidate phosphatase, *Biochim. Biophys Acta*, 56, 365, 1962.

66. **Agranoff, B. W.**, Hydrolysis of long-chain alkyl phosphates and phosphatidic acid by an enzyme purified from pig brain, *J. Lipid Res.*, 3, 190, 1962.

67. **Hokin, L. E., Hokin, M. R., and Mathison, D.**, Phosphatidic acid phosphatase in the erythrocytes membrane, *Biochim. Biophys. Acta*, 67, 485, 1963.

68. **Coleman, R. and Hübscher, G.**, Metabolism of phospholipids. V. Studies of phosphatidic acid phosphatase, *Biochim. Biophys. Acta*, 56, 479, 1962.

69. **McCaman, R. F., Smith, M., and Cook, K.**, Intermediary metabolism of phospholipids in brain tissue, *J. Biol. Chem.*, 240, 3513, 1965.

70. **Brindley, D. N., Smith, M. E., Sedgwick, B., and Hübscher, G.**, The effect of unsaturated fatty acids and the particle-free supernatant on the incorporation of palmitate into glycerides, *Biochim. Biophys. Acta*, 144, 285, 1967.

71. **Mitchell, M. P., Brindley, D. N., and Hübscher, G.**, Properties of phosphatidate phosphohydrolase, *Eur. J. Biochem.*, 18, 214, 1971.

72. **Åkesson, B., Elovson, J., and Arvidson, G.**, Initial incorporation into rat liver glycerolipids of intraportally injected [³H]glycerol, *Biochim. Biophys. Acta,* 210, 15, 1970.
73. **Åkesson, B., Elovson, J., and Arvidson, G.**, Initial incorporation into rat liver glycerolipids of intraportally injected [9,10-³H₂]palmitic acid, *Biochim. Biophys. Acta,* 218, 44, 1970.
74. **Rosenthal, A. F. and Pousada, M.**, Inhibition of particulate phosphatidate phosphohydrolase by an analogue of its substrate, *Biochim. Biophys. Acta,* 125, 265, 1966.
75. **Smith, M. E., Sedgwick, B., Brindley, D. N., and Hübscher, G.**, The role of phosphatidate phosphohydrolase in glyceride biosynthesis, *Eur. J. Biochem.,* 3, 70, 1967.
76. **Hübscher, G.**, Glyceride metabolism, in *Lipid Metabolism,* Wakil, S. J., Ed., Academic Press, New York, 1970, chap. VII.
77. **Casola, P. G., Macdonald, P. M., McMurray, W. C. C., and Possmayer, F.**, Concerning the coidentity of phosphatidic acid phosphohydrolase and phosphatidylglycerophosphate phosphohydrolase in rat lung lamellar bodies, *Exp. Lung Res.,* 3, 1, 1982.
78. **Johnston, J. M., Reynolds, G., Wylie, M. B., and Macdonald, P. C.**, The phosphatidate phosphohydrolase activity in lamellar bodies and its relationship to phosphatidylglycerol and lung surfactant formation, *Biochim. Biophys. Acta,* 531, 65, 1978.
79. **Casola, P. G. and Possmayer, F.**, Pulmonary phosphatidic acid phosphohydrolase: further studies on the activities in rat lung responsible for the hydrolysis of membrane-bound and aqueous dispersed phosphatidate, *Can. J. Biochem.,* 59, 500, 1981.
80. **Aurebekk, B. and Little, C.**, Function of arginine in phospholipase C of *Bacillus cereus, Int. J. Biochem.,* 8, 757, 1977.
81. **Coleman, R.**, Phosphatidate phosphohydrolase activity in liver cell surface membranes, *Biochim. Biophys. Acta,* 163, 111, 1968.
82. **Sedgwick, B. and Hübscher, G.**, Metabolism of phospholipids. IX. Phosphatidate phosphohydrolase activity in rat liver, *Biochim. Biophys. Acta,* 106, 63, 1965.
83. **Wilgram, G. F. and Kennedy, E. P.**, Intracellular distribution of some enzymes catalysing reactions in the biosynthesis of complex lipids, *J. Biol. Chem.,* 238, 2615, 1963.
84. **Tietz, D. F. and Shapiro, B.**, The synthesis of glycerides in liver homogenates, *Biochim. Biophys. Acta,* 19, 374, 1956.
85. **Stein, Y., Tietz, A., and Shapiro, B.**, Glyceride synthesis by rat liver mitochondria, *Biochim. Biophys. Acta,* 26, 286, 1957.
86. **Brindley, D. N. and Hübscher, G.**, The intracellular distribution of the enzymes catalysing the biosynthesis of glycerides in the intestinal mucosa, *Biochim. Biophys. Acta,* 106, 495, 1965.
87. **Smith, M. E. and Hübscher, G.**, The biosynthesis of glycerides by mitochondria from rat liver, *Biochem. J.,* 101, 308, 1966.
88. **Johnston, J. M., Rao, G. A., Lowe, P. A., and Schwartz, B. E.**, The nature of the stimulatory role of the supernatant fraction on triglyceride synthesis by the alpha-glycerophosphate pathway, *Lipids,* 2, 14, 1967.
89. **Moller, F. and Hough, M. R.**, Effect of salts on membrane binding and activity of adipocyte phosphatidate phosphohydrolase, *Biochim. Biophys. Acta,* 711, 521, 1982.
90. **Brindley, D. N.**, The intracellular phase of fat absorption, in *Biomembranes,* Vol. 4B, Smyth, D. H., Ed., Plenum Press, New York, 1974, chap. 12.
91. **Wilson, J. E.**, Brain hexokinase, the prototype ambiquitous enzyme, *Curr. Top. Cell. Regul.,* 16, 1, 1980.
92. **Moller, F., Wong, K. H., and Green, P.**, Control of fat cell phosphatidate phosphohydrolase by lipolytic agents, *Can. J. Biochem.,* 59, 9, 1981.
93. **Brindley, D. N.**, Intracellular translocation of phosphatidate phosphohydrolase and its possible role in the control of glycerolipid synthesis, *Prog. Lipid Res.,* 23, 115, 1984.
94. **Martin, A., Hales, P., and Brindley, D. N.**, A rapid assay for measuring the phosphatidate phosphohydrolase in cytosolic and microsomal fractions of rat liver, *Biochem. J.,* 245, 347, 1987.
95. **Martin, A., Hales, P., and Brindley, D. N.**, unpublished data.

Chapter 2

PHOSPHATIDATE PHOSPHOHYDROLASE ACTIVITY IN THE LIVER

David N. Brindley

TABLE OF CONTENTS

I. INTRODUCTION

In order to appreciate the control and function of phosphatidate phosphohydrolase in the liver it is necessary to consider the types of glycerolipids that are synthesized in this organ and the reasons why they are produced. As with most cells the liver needs to turn over its membrane lipids including phosphatidylcholine and phosphatidylethanolamine. These lipids are derived from diacylglycerol which in turn is produced mainly from phosphatidate (Chapter 1, Figure 1). The requirement for membrane synthesis is particularly important during growth and liver development, or when the liver is regenerating after partial hepatectomy.

In addition to these functions which are common to most cells, the liver is also required to secrete phosphatidylcholine in the form of a mixed micelle with cholesterol into the bile. The phosphatidylcholine helps to stabilize the micelle and it enables cholesterol to be transported in an aqueous environment. The biliary cholesterol is not completely reabsorbed from the small intestine and therefore the liver is able to promote the excretion of cholesterol from the body by this route. Furthermore, the phosphatidylcholine that enters the intestinal tract probably aids fat absorption. First, the phosphatidylcholine can be partially hydrolyzed to lysophosphatidylcholine which is a powerful detergent. This lipid facilitates the formation of micelles with bile salts and the fatty acids, monoacylglycerol, and cholesterol which results from the digestion of dietary fat. The formation of these mixed micelles is a prerequisite for efficient fat absorption.[1] Once in the enterocytes the lysophosphatidylcholine can be esterified back to phosphatidylcholine which is incorporated into the outer surface of chylomicrons. There, lipoproteins are responsible for transporting the majority of absorbed fat around the body.

A further possible function for the biliary phosphatidylcholine is to provide unsaturated fatty acids that can stimulate the absorption of saturated acids.[1] This may be particularly important in ruminant animals in which polyunsaturated acids in the diet are largely hydrogenated by bacterial action.

Apart from the synthesis of phospholipids, the liver is a major site in the body for the production of triacylglycerols. This synthesis enables fatty acids to be stored temporarily in the liver. It removes the potentially toxic effects of high concentrations of fatty acids and their coenzyme (CoA) esters and it also regenerates the CoA. The triacylglycerol can subsequently be hydrolyzed and used for β-oxidation, or alternatively, the triacylglycerol can be secreted in very low density lipoproteins (VLDL). This latter process enables the liver to transport energy to other organs such as adipose tissue and muscle. VLDL contain about 60% of triacylglycerol and about 20% phosphatidylcholine and so their secretion requires the coordinated production of these two lipids.

Table 1
**INCREASES IN THE ACTIVITY OF PHOSPHATIDATE
PHOSPHOHYDROLASE IN THE LIVER IN SOME CONDITIONS
IN WHICH THE EFFECTS OF GLUCOCORTICOIDS AND
OTHER STRESS HORMONES ARE INCREASED RELATIVE TO
INSULIN**

Condition of treatment	Duration	Increase in phosphatidate phosphohydrolase activity	Ref.
A. Metabolic stress			
Starvation (rats)	6—40 hr	1.3—2.3-fold	16—18
Surgical stress (rats)	6 hr	3-fold	17
Subtotal hepatectomy (rats)	6 hr	5.5-fold	17
Diabetes (mildly ketotic in rats)	10 weeks	1.4-fold	19
Diabetes (ketotic in rats)	48—72 hr	2—4-fold reversed by insulin	20—22
Hypoxia (rats)	24 hr	2.5-fold	18
B. Toxic conditions caused by			
Hydrazine (rats)	4—24 hr	2—4-fold	23
Morphine (mice)	60—180 min	2-fold	25
C. Glucocorticoid effects in vivo			
Injection of corticotropin	4 hr	1.4-fold	26
Injection of corticotropin	4 hr	3.3—4.3-fold	27
Injection of cortisol	5 days	2.4-fold	8
Injection of cortisol	4 hr	1.9-fold	26

Finally, the liver also synthesizes some of the components of high density lipoproteins (HDL). Some of the phospholipids of these lipoproteins may be secreted with the apolipoproteins (apo C and apo A) or they may be transferred to the HDL from the coat material of chylomicrons and VLDL as these are degraded by the action of lipoprotein lipase.

II. PHYSIOLOGICAL REGULATION OF PHOSPHATIDATE PHOSPHOHYDROLASE IN THE LIVER

A change in the capacity of the liver to synthesize triacylglycerols appears to be normally associated with a parallel change in the total activity of phosphatidate phosphohydrolase. Some of these changes in activity are summarized in Tables 1 and 2 and they are generally far more dramatic than those observed in the activities of other enzymes involved in triacylglycerol synthesis.[3,4] In some cases, it is only the phosphohydrolase activity that changes at all. These results in themselves suggest that the phosphohydrolase is involved in the control of hepatic triacylglycerol synthesis. Further evidence to support this hypothesis comes from the observations that the accumulation of phosphatidate in the liver can vary inversely with respect to the estimated rate of triacylglycerol biosynthesis.[5-9] However, the extent of the accumulation of phosphatidate in the membranes of cells is likely to be limited by the phospholipases of the A type that have relatively high activities in the liver towards phosphatidate.[10-15] These activities, in fact, may help to regulate the direction of phosphatidate metabolism.[4]

Table 2
**ACUTE EFFECTS OF NUTRIENTS AND THE EFFECTS OF
CHRONIC DIETARY MODIFICATION ON THE ACTIVITY OF
PHOSPHATIDATE PHOSPHOHYDROLASE IN THE LIVER**

Treatment	Duration	Increase in phosphatidate phosphohydrolase activity	Ref.
Fructose, sorbitol, glycerol, or dihydroxyacetone	6—8 hr	1.6—2.3-fold	14, 31, 32
Ethanol (rats)	4—7 hr	2—7-fold	6, 14, 31, 32—35
Ethanol (hamster)	5—10 hr	2.3—3-fold	36
Ethanol (baboons)	1—7 years	1.2—1.5-fold	37
High (75% by weight) glucose or fructose in diet (rats)	60 hr	3-fold	42
Replacement of 20% (by weight) of starch by sucrose (rats)	14 days	No change	9
Replacement of 40% (by weight) of starch by sucrose, beef tallow, or corn oil (rats)	21 days	No change	27
Replacement of 20% (by weight) of starch by lard (rats)	14 days	1.3—1.4-fold	9
Mustard seed or corn oil at 20% by weight (rats)	1, 3, or 6 weeks	No change	46
Essential fatty acid deficiency (rats)	7 weeks	2.2—2.8-fold	47
(a) Lard supplemented diet followed by (b) acute feeding with glucose or ethanol (rats)	(a) 14 days (b) 6 hr	5-fold	56
Obesity (ob/ob mice)	Long term	1.8-fold	63, 64

A. Metabolic Stress, Starvation, Diabetes, and Toxic Conditions

A common feature of many of the conditions that lead to an increase in hepatic phosphatidate phosphohydrolase is "metabolic stress". This term is used here to express a state of hormonal balance in which the effects of the stress hormones, glucagon, catecholamines, corticotropin, and glucocorticoids tend to predominate relative to the action of insulin. This type of situation is seen in starvation, surgical stress, diabetes, hypoxia, and after subtotal hepatectomy, in which there are also significant increases in the activity of phosphatidate phosphohydrolase (Table 1). Many of these conditions are also accompanied by the accumulation of triacylglycerol in the liver. Part of this is probably caused by the increased supply of fatty acids to the liver as a result of increased lipolysis in adipose tissue. It is significant to note that the increase in the activity of phosphatidate phosphohydrolase in the liver in rats with ketotic diabetes can be reversed by the injection of insulin.[20,21] This fall in activity is paralleled by a decrease in the concentration of hepatic triacylglycerol. The injection of insulin will also decrease the release of fatty acids from adipose, and thus the subsequent activation of phosphatidate phosphohydrolase by these acids (see Section IV.B). In the longer term, insulin will also decrease the rate of synthesis of the phosphohydrolase and increase its rate of inactivation (see Sections III.A.4 and III.B).

It was also proposed that insulin may modify the ability of triacylglycerols to feed back and inhibit the activity of phosphatidate phosphohydrolase.[22] Triolein emulsions and the nascent form of very low density lipoproteins inhibited the synthesis of di- and triacylglycerols. This effect was shown to be at the level of phosphatidate phosphohydrolase since serum very low density lipoproteins inhibited the cytosolic activity of this enzyme in a dose-dependent manner. A high I_{50} value was found with the phosphohydrolase isolated from the livers of diabetic rats and the I_{50} value was decreased by treatment with insulin. It was concluded that this reduced sensitivity to inhibition by triacylglycerols may partly explain the increased rate of synthesis of triacylglycerols that was observed with liver homogenates from diabetic rats.[22]

The increase in the phosphohydrolase activity of the liver remnant after subtotal hepatectomy was particularly marked compared with the sham operated and starved controls.[17] Maximum activity was seen at 10 to 16 hr after the sham operation in the subtotal hepatectomy. This compares with a peak accumulation of triacylglycerols in the liver remnant at 16 hr after the operation (Figure 1). By 24 hr, the activity of the phosphohydrolase and the triacylglycerol content of the liver had begun to fall. In the sham-operated rats the phosphohydrolase activity had, in fact, reached that in the starved controls.

The injection of actinomycin D prevented the increase in the phosphohydrolase activity of rats that had been subjected to subtotal hepatectomy.[17] This, together with the time course of the response in the phosphohydrolase activity suggests that the increases were caused by the synthesis of new enzyme protein.

It can also be seen from the work of Mangiapane et al.[17] that there was relatively little change in the microsomal activities of palmitoyl CoA synthetase, glycerol phosphate acyltransferase, and diacylglycerol acyltransferase after subtotal hepatectomy and the sham operation. In the starved controls there was, in fact, a decrease in the activity of palmitoyl-CoA synthetase and glycerol phosphate acyltransferase during starvation at 16 to 24 hr.[17]

The increase in the phosphohydrolase activity that occurs in hypoxia[18] can also be understood in terms of the ''stress reaction'' that will occur in this condition. Furthermore, the clearance of fatty acids by β-oxidation will be hampered by the lack of oxygen and therefore the liver needs to be able to remove the excess fatty acids and acyl-CoA esters by sequestering them as triacylglycerols.

A variety of toxic conditions can lead to a severe stress reaction and the mobilization of fatty acids from adipose tissue. This is again accompanied by an increased activity of phosphatidate phosphohydrolase as indicated by the effects of hydrazine,[23,24] morphine,[25] and ethanol (Table 1). In the case of hydrazine[24] and ethanol (Figure 2) there is a marked increase in the circulatory concentrations of glucocorticoids relative to insulin. The time course for the increase in phosphohydrolase after administration of hydrazine and ethanol is compatible with an induction of enzyme synthesis. However, in the case of morphine, the increase in the phosphohydrolase activity and in the accumulation of triacylglycerol in the liver was seen after the relatively short time of 2 hr. Part of the increase should have been brought about by an induction of enzyme synthesis since the effects of morphine were decreased by cycloheximide. However, the exposure of monolayer cultures of hepatocytes or of microsomal fractions from the liver to morphine also produces an increase in triacylglycerol production and in the phosphohydrolase activity over periods of 30 to 120 min.[25] This suggests a direct effect of morphine in addition to an effect through stimulating the release of corticotropin and glucocorticoids. A further explanation of the morphine effects is that the rise in phosphatidate phosphohydrolase activity was measured in the microsomal fraction. This could have been partly produced by a translocation of the cytosolic activity onto the membranes of the endoplasmic reticulum (see Section IV.B).

The work described in this section provides evidence that an increased control of metabolism by the stress hormones relative to insulin is accompanied by large increases in the hepatic activity of the phosphohydrolase. This provides the liver with an increased capacity

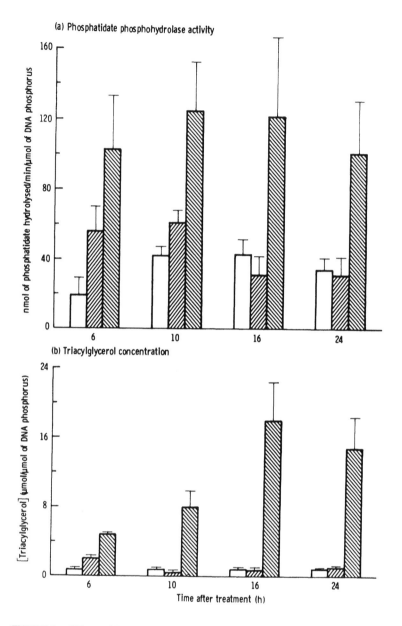

FIGURE 1.   Effects of food deprivation, sham operation, and subtotal hepatectomy on the activity of phosphatidate phosphohydrolase and on the concentration of triacylglycerol in the liver. The figure shows the activity of the phosphohydrolase and the concentration of triacylglycerol as a function of time after the sham operation ▨, or after subtotal hepatectomy ▧. Control rats were deprived of food ☐. Results are mean ± SD.[17]

for esterifying fatty acids and for the synthesis of triacylglycerols. However, the expression of this capacity depends upon a variety of factors, the most important of which is probably the net rate of fatty acid supply, taking into account changes in β-oxidation. For example, the increased activity in starvation need not be expressed because the increased fatty acid supply to the liver from adipose tissue can be diverted into β-oxidation. If this latter process is inhibited by octanoyl-(+)-carnitine, then the fatty acids are readily incorporated into triacylglycerols, provided that glycerol phosphate concentrations are optimum.[28,29] The concentrations of glycerol phosphate and of long chain fatty acids and their CoA esters can be

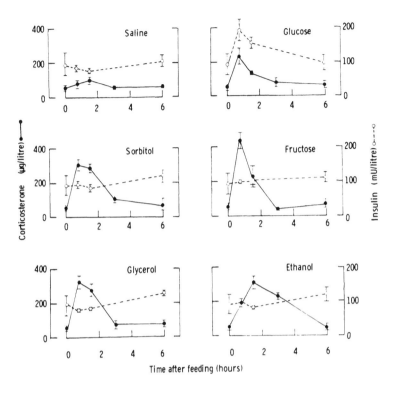

FIGURE 2. Effects of feeding rats with glucose, sorbitol, fructose, glycerol, and ethanol on the concentrations of insulin and corticosterone in their blood sera. Rats were fed by stomach tube[14] with various nutrients that are indicated and the concentrations of insulin (○) and corticosterone (●) were measured in the blood sera at various times after feedings. Results are means ± SEM for 4 rats per group except for the glucose group at 45 min, the ethanol group at 4 hr, and for the untreated rats for which there were 8, 3, and 10 rats, respectively. (From Brindley, D. N. et al., *Biochem. J.*, 180, 195, 1979. With permission.)

maintained or increased in vivo by the mobilization of glycerol and fatty acids from adipose tissue in starvation.[29,30] The adaptive increase in the phosphohydrolase activity provides the liver with the ability to protect itself against the toxic accumulation of the fatty acids and their CoA esters.

B. Dietary Modification

Feeding rats by stomach tube or injection with fructose, sorbitol, glycerol, dihydroxy-acetone, or ethanol causes an increase in the activity of phosphatidate phosphohydrolase present in the liver when compared with control rates treated with saline or glucose (Table 2). The increases took place over a period of hours. The increases were reflected in the soluble and microsomal activities of the enzyme. The activities of other enzymes of glycerolipid synthesis, including glycerol phosphate acyltransferase and diacylglycerol acyltransferase in rat liver were not increased by an acute dose of ethanol to a significant extent.[33]

The administration of acute loads of fructose, sorbitol, glycerol, and ethanol to rats produces a marked increase in the concentration of circulating corticosterone. By contrast, the concentrations of insulin do not change significantly (Figure 2). Feeding the rats with an equal energy load of glucose produced a small rise in the concentration of circulating corticosterone and simultaneously stimulated the secretion of insulin. In this case there was no significant change in the phosphohydrolase activity in the liver. These results also support the hypothesis that the stimulus for increasing the phosphohydrolase activity in vivo is an increase in the ratio of circulating glucocorticoid to insulin.

Adrenalectomized rats that had been maintained by providing saline in their drinking water only showed a 1.7-fold increase in the phosphohydrolase activity compared with a 6.9-fold increase after 7 hr in normal rats.[34] This result demonstrates that a large proportion of the increase is probably caused by the release of glucocorticoids (see Section III.A.1).

The residual increase of 1.7-fold in the phosphohydrolase activity that was observed in the livers of the adrenalectomized rats[32] could have been caused by the stimulations that can be produced by fatty acids (Section IV.B) and glucagon (Section III.A.2).

Alternatively, it has been suggested that increases in the concentration of glycerol phosphate and the increases in the [NADH]/[NAD] ratio (redox state) that are produced by ethanol could be important in controlling the phosphohydrolase activity.[31,33] This hypothesis is supported by the observation that pyrazole, which inhibits ethanol oxidation, also prevents part of the ethanol-induced increase in the soluble but not the membrane-bound phosphohydrolase activity.[35] The mechanisms by which the redox state of the liver could change the phosphohydrolase activity are not known but changes in the rate of phosphohydrolase synthesis or degradation may be involved over the 12-hr period that was employed.[35] A direct effect of redox state in controlling the phosphohydrolase activity has also been demonstrated with hepatocytes in culture in which pyrazole partially prevented the increase.[38] This involvement of the redox state can only account for a small portion of the changes in the phosphohydrolase activity since adrenalectomy seemed to abolish about 85% of the ethanol-induced increase in the phosphohydrolase activity.[32] It is also relevant to note that feeding fructose, which should not alter the hepatic redox state, increased the soluble phosphohydrolase activity to the same extent as did sorbitol. The latter compound could increase the [NADH]/[NAD] ratio.[32] It is also relevant to note that Hassinen et al.[34] subsequently treated rats in vivo with pyrazole and observed that this did not prevent the ethanol-induced increase in the phosphohydrolase activity after 4 hr at concentrations where pyrazole did prevent the changes in the redox state.

Further evidence for the major involvement of glucocorticoids in producing the increases in hepatic phosphatidate phosphohydrolase activity after dosing rats acutely with ethanol is provided by work with the hypotriglycerdemic drug, benfluorex (see Section II.H). Chronic treatment of rats with this compound diminishes the ethanol-induced increase in circulating corticosterone and it partially prevents the rise in phosphohydrolase activity.[39,40]

A slightly different experimental approach that has been used to investigate the effects of glucocorticoids in vivo involves adrenalectomized rats that were then maintained by injections of cortisol or corticosterone.[35] The metabolism of the liver can therefore be influenced by basal concentrations of glucocorticoids, whereas the subsequent administration of ethanol will not stimulate the secretion of additional glucocorticoid. In these adrenalectomized rats, ethanol produced a 1.7-fold increase in the activity of the soluble phosphatidate phosphohydrolase compared with a 1.6-fold increase in the control rats.[35] However, these increases are relatively modest in scale compared with those reported in other studies.[32] Although adrenalectomy followed by glucocorticoid therapy did not appear to modify significantly the ethanol-induced increase in the soluble phosphohydrolase activity it did prevent the rise in the microsomal activity.[35] It is this membrane-bound activity that is thought to be the active form of the enzyme, and the movement of the soluble phosphohydrolase to the endoplasmic reticulum is thought to be mainly controlled by the accumulation of fatty acids and their CoA esters in the liver.[41]

So far the effects of ethanol that have been discussed have resulted from a single dose of ethanol. Chronic ethanol administration to hamsters produced an increase in the phosphohydrolase activity that was sustained for at least 6 weeks.[36] The administration of a liquid diet which contained 50% of its energy in the form of ethanol for 1 to 7 years to baboons also increased the hepatic activity of phosphatidate phosphohydrolase.[37] This was accompanied by an accumulation of triacylglycerol in the liver and a hypertriglyceridemia. In the early stages where there was a fatty liver without perivascular fibrosis, there was an increase in the activities of diacylglycerol acyltransferase in the microsomal fraction and of phosphatidate phosphohydrolase in the soluble and microsomal fractions. As the damage to the liver progressed with the development of septal fibrosis and/or cirrhosis, the rate of triacylglycerol accumulation in the liver and the hypertriglyceridemia decreases. This was accompanied by decreases in the activities of diacylglycerol acyltransferase and the soluble phosphatidate phosphohydrolase.[37] The microsomal phosphatidate phosphohydrolase activity did not decrease significantly and the glycerol phosphate acyltransferase activity was not significantly higher than in the controls. These results also suggest that increases in the synthesis of triacylglycerol in the liver are facilitated by changes in the activities of phosphatidate phosphohydrolase and diacylglycerol acyltransferase rather than by glycerol phosphate acyltransferase. However, direct measurements of triacylglycerol synthesis were not made in these experiments.[37] The relationship between the microsomal and cytosolic phosphohydrolase activities now needs to be considered in light of the translocation of the enzyme between these compartments (Section IV.B).

Persistent increases in the hepatic activity of phosphatidate phosphohydrolase were also demonstrated[42] when rats were fed artificial diets that were enriched with glucose or fructose (Table 2). High fructose diets when fed to rats for 11 days also increased the synthesis of neutral lipid from glycerol phosphate.[5] The enhancement of the phosphohydrolase activity was also indicated in these experiments by the fall in the concentration of phosphatidate in the microsomal fraction and the increase in the diacylglycerol concentration. The activity of diacylglycerol acyltransferase was increased in these investigations in which the assay was performed with membrane-bound diacylglycerol.[5]

It is not certain whether this chronic effect of dietary fructose[42] on the phosphohydrolase activity was caused by the increased release of glucocorticoids as was the case with rats that were fed acute loads of fructose.[32] There is some evidence that diets rich in sucrose can cause this to happen.[43,44] However, feeding 20% by weight of sucrose to rats in artificial diets for 14 days produced no significant change in the phosphatidate phosphohydrolase activity in the cytosol of the liver.[9] Since glucocorticoids and other stress hormones are known to be so important in regulating the activity of the phosphohydrolase, it is important to assess the effect of the physiological and psychological stress that is caused to experimental animals when their diets are modified (see below). These factors have to be dissociated from the metabolic changes that are caused by the removal or addition of specific nutrients. In experiments where rats were fed for 3 weeks on pelleted diets that contained 40% by weight of sucrose rather than starch, there was also no change in the soluble phosphatidate phosphohydrolase activity in the liver.[27] Care was taken that these diets were nutritionally balanced and the rate of growth of the different groups of rats were very similar. Sucrose feeding did, however, increase the circulating concentration of triacylglycerol[27] and this might have been partly caused by the fall in lipoprotein lipase activity in the adipose tissue.[45]

The type and quantity of fat that are present in the diet can modify the rate of hepatic triacylglycerol synthesis. Rats that were fed on powdered diets enriched with lard synthesized more triacylglycerol than did rats fed a starch diet and they had a slightly higher phosphohydrolase activity in their livers.[9] However, in subsequent work pelleted diets enriched with beef tallow or corn oil were used and their effects compared to those of a starch-rich diet.[27] The energy obtained from protein relative to the total energy content was approximately the same in these diets. The basal activity of phosphatidate phosphohydrolase in the livers of the rats fed these three different diets was not significantly different. It has also been shown that the addition of mustard-seed oil or corn oil to the diets of rats did not alter the phosphohydrolase activity in the liver.[46] However, this activity in microsomal and cytosolic fractions did increase in essential fatty acid deficiency.[47] This could possibly have been caused by an increased control of metabolism by the stress hormones in these animals.

Animals that are fed on high fat diets often exhibit insulin insensitivity[48-51] with a decreased number of insulin-binding sites on adipocytes and hepatocytes[52,53] and a magnified response to various stress stimuli.[54,55] This is also reflected in a prolonged release of corticosterone in response to feeding an acute dietary load of fructose.[56] These rats were fed diets enriched with beef tallow or corn oil and the corticosterone response was particularly enhanced with the latter diet. These diets together with the control diets that were rich in carbohydrate were carefully balanced nutritionally and they were fed in pellet form. The growth rates of the rats were rapid and similar on the different diets.[27,56]

The stress responses in terms of corticosterone release to fructose feeding were even more pronounced and prolonged in rats that were fed artificial powdered diets that were enriched with corn oil or lard.[56] These rats had poor growth rates and they even exhibited large corticosterone responses after glucose feeding. This was accompanied by an increase of about threefold in the activity of the cytosolic phosphatidate phosphohydrolase in the liver.[56] By contrast, the corticosterone response to glucose feeding was much smaller in rats maintained on the normal 41B diet[32] that was also rich in starch and it was balanced by a relatively large secretion of insulin. In this case the soluble phosphohydrolase activity was unchanged after 6 hr.[32]

It was concluded from this work that the artificial powdered diets increased the susceptibility of the rats to a subsequent stress stimulus such as the acute feeding of a glucose or fructose load. This event was particularly apparent when the diets were enriched with fat. This effect of dietary fat was also observed in pellet diets which were well tolerated by the rats.[56] The exaggerated and prolonged release of corticosterone leads to a more prolonged increase in phosphatidate phosphohydrolase and in the potential for synthesis of triacylglycerols within the liver.[56] These observations could partly explain why the effects of dietary sucrose (fructose),[57,58] sorbitol,[59] or ethanol[60-62] in stimulating hepatic triacylglycerol production are exaggerated in animals fed high fat diets. In addition, dietary fat is a major contribution to the fatty acid that accumulated in liver after ethanol consumption.

The activity of phosphatidate phosphohydrolase is also increased in the livers of obese mice (ob/ob)[63,64] and this probably relates both to the overnutrition of these animals and to the high circulating concentrations of corticosterone.[65] Measurements of phosphatidate phosphohydrolase from the cytosol of the livers of obese patients also appeared to show a slightly increased activity but this did not reach the level of statistical significance. This work is discussed more fully in Section IV.E.

C. Effects of Glucocorticoids and Corticotropin

The injection of rats for 5 days with cortisol significantly increased the activity of phosphatidate phosphohydrolase in the liver (Table 1). This increase was accompanied by an increased rate of synthesis of triacylglycerols in the liver which was measured 1 min after the intraportal injection of [3H]-glycerol and [14C]-palmitate. The role of the phosphohydrolase in facilitating these increases was emphasized by the decreased accumulation of 3H and 14C in the phosphatidate.[8]

A shorter-term treatment of rats with glucocorticoids or corticotropin is also effective in increasing the phosphohydrolase activity. For example, rats that were injected with cortisol[26] and corticotropin (Synacthen)[26,27] showed an increase in the phosphohydrolase activity in the liver after 4 or 6 hr.

The treatment of animals with glucocorticoid or corticotropin is known to cause increases in (1) the synthesis of triacylglycerols in the liver,[8] (2) the concentrations of triacylglycerol in the liver,[66,67] and (3) the secretion of very low density lipoproteins.[68-72] The increased phosphohydrolase activity probably facilitates the increased triacylglycerol synthesis,[8] although a direct relationship between these events and the secretion of triacylglycerols was not demonstrated with cultured hepatocytes.[72] This will discussed further in Section VI.

D. Effects of L-Thyroxine

The administration of L-thyroxine to rats for 5 days increased the ability of liver homogenates to synthesize diacylglycerols and triacylglycerols from glycerol phosphate but without significantly increasing the activity of glycerol phosphate acyltransferase.[73] These results indicate an increased activity of phosphatidate phosphohydrolase. This conclusion was later confirmed in rats that had been injected for 5 or 7 days with L-thyroxine when the phosphohydrolase activity was expressed per milligram of soluble protein[8] or per gram wet weight,[31] respectively. However, there was no significant increase in the phosphohydrolase activity when the results were expressed per total liver. The long-term treatment of rats of L-thyroxine also stimulated the synthesis of triacylglycerols as measured in vivo from intraportally injected [^{3}H]-glycerol and [^{14}C]-palmitate.[8] This may be part of a general stimulation of metabolism. The increase in triacylglycerol synthesis was also accompanied by a decrease in the accumulation of ^{3}H and ^{14}C in diacylglycerol.[8] This possibly indicates a stimulation in diacylglycerol acyltransferase activity as suggested from earlier work.[74]

The effect of thyroxine on phosphatidate phosphohydrolase activity in the liver was not seen 4 hr after its injection. This is consistent with some of the other metabolic effects of thyroxine which may take many days to become apparent.[26]

E. Effects of Sex and Sex Steroids

It is generally observed that the livers of female animals are able to synthesize and secrete triacylglycerol at a greater rate than those of males[75-78] and that the female livers have a higher triacylglycerol concentration.[79-81] The activity of the soluble and microsomal phosphatidate phosphohydrolases was found by Savolainen et al.[82] to be higher in the livers of fed adult female rats than in the males. This was also accompanied by a higher triacylglycerol concentration. By contrast, Goldberg et al.[83] reported relatively little difference between the microsomal or soluble phosphohydrolase activities of fasted male and female rats. These authors did, however, show that there was a greater proportion of the phosphohydrolase on the membranes relative to that in the cytosol.[83] If the membrane-bound enzyme is taken to be the active form[41] then this is again compatible with an increased ability of the female liver to synthesize triacylglycerols and the higher concentration of triacylglycerol that was observed in the liver and in the serum.[83] However, the concentration of hepatic triacylglycerol as a function of the age of the female rats appears to be correlated with the soluble rather than the microsomal phosphohydrolase activities.[83] It is now known that the soluble phosphohydrolase can reversibly translocate onto the microsomal membranes in response to the presence of fatty acids, acyl-CoA esters, and phosphatidate in the membranes (Section IV.B). The distribution that exists within the cell in vivo may not be entirely maintained from the time when the animal is killed to when the cell fractions are prepared.

A variety of experiments have been performed to investigate the effects of sex hormones on the phosphohydrolase activity. Gonadectomy of male rats caused increases of 25 and 80% respectively after 6 weeks in the soluble and microsomal phosphohydrolase activities.[82] By contrast, gonadectomy of the female failed to significantly alter the microsomal or soluble phosphohydrolase activities. The concentration of triacylglycerol in the livers of the male and female rats reflected the relative activities of the phosphohydrolase.

Gonadectomy of the male rats did not significantly alter the activities of the microsomal and soluble phosphohydrolase after 2 weeks in contrast to the increases that were observed after 6 weeks.[82] Implants of testosterone or estradiol on the same day as the gonadectomy did not alter either of the phosphohydrolase activities in the liver of male rats after 14 days.

In other work on intact rats the injection of 5 mg estradiol per kilogram or equimolar doses of 6α-methyl-17α-hydroxyprogesterone acetate or testosterone to male or female rats for 5 days did not significantly alter the phosphohydrolase activity in the liver.[227] Experiments with cultures of rat hepatocytes have so far failed to demonstrate effects of testosterone,

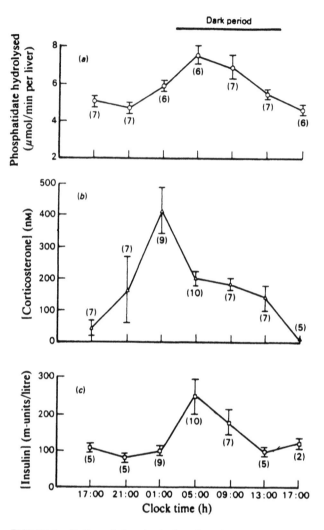

FIGURE 3.   Daily variations in the hepatic activity of the soluble phosphatidate phosphohydrolase and in the serum concentrations of corticosterone and insulin in the male rats. The results are means ± SEM and the numbers of rats are indicated in parentheses. (From Knox, A. M. et al., *Biochem. J.*, 180, 441, 1979. With permission.)

estrogens, or progestogens within the physiological range on the activity of phosphatidate phosphohydrolase.[228] Concentrations of progesterone in the micromolar range did increase the total phosphohydrolase activity of the hepatocytes but this was probably caused by the interaction of this high concentration of progesterone with the glucocorticoid receptor.

It is concluded from the work described in this section that although differences can be found between the metabolism of triacylglycerols and the activity of phosphatidate phosphohydrolase activity in the livers of male and female rats, a direct action of sex steroids on the phosphohydrolase activity has yet to be demonstrated.

F. Circadian Rhythm

The activity of the soluble phosphatidate phosphohydrolase activity in the livers of normally fed rats reached a peak at about 2 hr after the beginning of the dark period.[84] This is the time when the rats are feeding and it coincides with the peak in the concentration of circulating insulin (Figure 3). The highest circulating concentrations of corticosterone occurred 2 hr before the beginning of the dark period. It is thought that it is this rise in glucocorticoid concentration when insulin concentrations are low that stimulates the synthesis of the phosphohydrolase. This is seen as a peak in this activity 4 hr later. By this time the corticosterone concentration has fallen and the insulin concentration in the blood has risen. This inhibits the synthesis of the phosphohydrolase and its activity subsequently falls. The justification for the interpretation is given in Section III.A.

The coincidence of the high phosphohydrolase activity at a time when the rat is feeding on its high carbohydrate diet possibly facilitates the synthesis of triacylglycerols in the liver and the subsequent secretion of very low density lipoproteins.

G. Effects of Age

Jamdar et al.[85] measured the rate of glycerolipid synthesis from glycerol phosphate in liver homogenates and in microsomal fractions from rats of different ages. The results from male and female animals were pooled. The rate of glycerolipid synthesis was fairly high in the liver from 20-day-old fetuses and it reached a maximum 1 day after birth. The total rate of glycerolipid synthesis then declined up to 10 days of age. There was a second increase between 15 to 21 days followed by a decline to adult levels. The increase in glycerol phosphate acyltransferase on the first day was prevented by the administration of puromycin which indicates a requirement for protein synthesis. Similar effects had been demonstrated with puromycin on the increases of other microsomal enzymes including cytochrome c reductase and glucose-6-phosphatase. The increase in the glycerol phosphate acyltransferase activity after birth was also prevented if the rats were not allowed to suckle. Although the mechanism for this latter effect was unclear, it was suggested that the normal increase in the nonesterified fatty acid content of serum and the liver, that occurs as a result of milk intake, may be particularly important. The total rate of glycerolipid synthesis did not appear to be controlled by acyl-CoA synthetase which had the highest specific activity in adult rats.[85]

The ratio of neutral lipid (diacylglycerol and triacylglycerol) relative to phospholipid (lysophosphatidate and phosphatidate) that was synthesized by the homogenates was high for the fetal and 1-day-old rats. This indicates a relatively high activity of phosphatidate phosphohydrolase. By contrast, the microsomal fractions obtained from rats of various ages synthesized little neutral lipid. This can be explained by the removal of the cystolic phosphatidate phosphohydrolase.

Coleman and Haynes[86] also demonstrated an increase in the capacity of the liver to synthesize glycerolipids just before birth and during the days immediately afterwards. The specific activities of acyl-CoA synthetase, lysophosphatidate acyltransferase, diacylglycerol cholinephosphotransferase, and diacylglycerol acyltransferase increased by 1.4- to 3.4-fold in the 4 days before birth. After birth, lysophosphatidate acyltransferase, choline phosphotransferase, and acyl-CoA synthetase increased by five- to tenfold by the eighth day post partum. The microsomal esterification of glycerol phosphate and dihydroxyacetone phosphate was undetectable until 3 days before birth and both activities increased by about 74-fold by the fifth day after birth.

These authors[86] reported relatively little change in the activity of phosphatidate phosphohydrolase in the microsomal or cytosolic fraction. However, they did question whether their assay system might not specifically detect the activity of the phosphohydrolase that was concerned with glycerolipid synthesis. In particular, they used phosphatidate derived from egg phosphatidylcholine but they do not state whether Ca^{2+} was removed before it was employed as a substrate (Section II.B). Secondly, the microsomal phosphohydrolase activity was determined by measuring the release of inorganic phosphate. This can also be formed by the action of A-type phospholipases followed by the hydrolysis of the glycerol phosphate by acid or alkaline phosphatases. Consequently, this method for determining the activity of phosphatidate phosphohydrolase in microsomal fractions from rat liver can be invalid.[13]

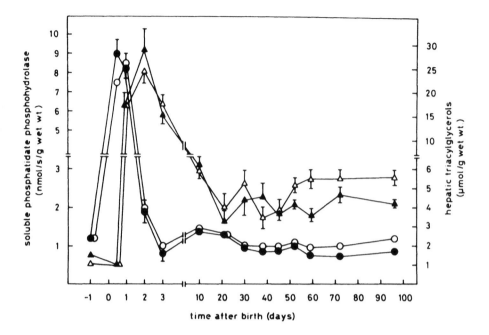

FIGURE 4.   Effect of age on soluble phosphatidate phosphohydrolase activity and triacylglycerol concentration in livers of feeding male and female rats. Each point represents mean ± SD for 3 to 7 animals. Symbols: —●—●—, phosphatidate phosphohydrolase activity, male; —○—○—, phosphatidate phosphohydrolase activity, female; —▲—▲—, triacylglycerol concentration, male; —△—△—, triacylglycerol concentration, female. (From Savolainen, M. J. et al., *Metabolism*, 30, 706, 1981. With permission.)

In other experiments on development the activity of phosphatidate phosphohydrolase was measured directly by determining the production of phosphate or diacylglycerol for the soluble and microsomal fractions respectively.[82] The phosphatidate that was used as a substrate was converted to the Na$^+$ form by using Chelex resin. Savolainen et al.[82] demonstrated a sevenfold increase in the activity of the soluble phosphatidate phosphohydrolase in the liver 12 hr after birth (Figure 4). This activity decreased to adult levels by the third day post partum. The changes in phosphohydrolase activity were followed by similar alterations in the concentration of hepatic triacylglycerol. The patterns of these changes were fairly similar in male and female rats except that the soluble phosphohydrolase activity was 25% higher in the livers of adult females and the hepatic triacylglycerol concentration was 20 to 50% greater (see also Section II.E).

Goldberg et al.[83] measured changes in both the microsomal and soluble phosphohydrolase activities with age and obtained slightly different results from those of Savolainen et al.[82] The initial changes on the first day after birth were not measured in the former studies. The activities of phosphatidate phosphohydrolase in the microsomal fractions showed only minor fluctuations with age in both male and female rats (Figure 5). By contrast the activity in the cytosolic fraction of the male rats rose to a peak at day 58 and then it declined. In females, the cytosolic activities were lower than in the males up to day 35 but there was then an increase in activity of about fivefold by day 50. From then on the activities in the two sexes were similar.[83]

The concentrations of triacylglycerol in the livers of the male rats remained relatively constant during the period of investigation.[83] In the females the triacylglycerol concentration of the liver approximately doubled from day 25 to day 50 and from then it remained higher than in the males. It was claimed that a fairly close relationship existed between the soluble phosphohydrolase activity and the hepatic triacylglycerol concentration in female rats. However, such a relationship was not apparent in male rats.[83]

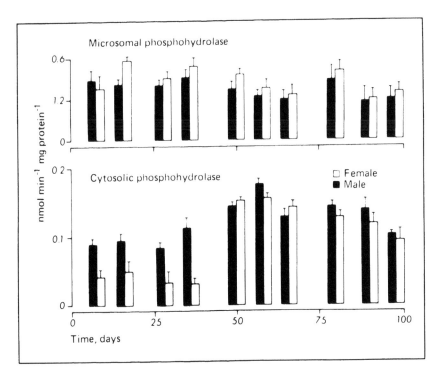

FIGURE 5. Relationship of microsomal (top) and cytosolic (bottom) phosphatidate phospho-hydrolase activity to age in the livers of fasting male and female rats. Each bar is mean value for 3 animals ± SEM. (From Goldberg, D. M. et al., *Enzyme*, 30, 59, 1983. With permission.)

It is relevant to note that the increased capacity of the neonatal rat to synthesize glycerolipid may not be entirely dependent on the synthesis *de novo* of phosphatidate and its conversion via diacylglycerol to triacylglycerol and phospholipids. The livers of suckling rats also possess a monoacylglycerol acyltransferase activity that is about 700-fold higher than adult livers.[87] This activity falls rapidly after about the eighth day post partum.

Monoacylglycerol acyltransferase is normally associated with the re-esterification of dietary triacylglycerol that is absorbed mainly in the form of monoacylglycerol by the enterocytes.[1,2] The enzyme that is located in the liver appears to be a different isoenzyme form compared to that found in the small intestine.[88]

During the suckling period the liver would be receiving large amounts of triacylglycerol from the chylomicron remnants derived from the transport of milk fat. This dietary fat provides the major source of energy for the neonate. The hepatic monoacylglycerol acyltransferase could therefore provide additional diacylglycerol for glycerolipid synthesis and thus augment the activity of phosphatidate phosphohydrolase in the suckling period.

H. Effects of Fenfluramine, Benfluorex, and Other Amphiphilic Amines

1. Direct Effects

Fenfluramine was introduced as an anorectic durg with which to replace the amphetamines in the treatment of obesity. The problem with amphetamines was its stimulatory properties which lead to drug abuse. The fenfluramine type of compounds did not have these effects. It was later shown that fenfluramine had other beneficial therapeutic effects that were independent of its action in decreasing food intake. These included its ability to decrease hyperglycemia and hypertriglyceridemia and to improve insulin sensitivity.[90,91] Benfluorex is a derivative of fenfluramine that was specifically introduced as a hypotriglyceridemic agent.[91]

Table 3

EFFECTS OF VARIOUS DRUGS AND OTHER COMPOUNDS ON THE ENZYMES OF GLYCEROLIPID SYNTHESIS

Concentration (mM) required to produce a 50% inhibition in the activity of

Compound	Glycerol phosphate incorporation into		Phosphatidate phosphohydrolase	Diacylglycerol acyltransferase
	Total lipids	Glycerides		
p-Chlorobenzoate	~80 (6)	~50 (6)	20 (8)*	n.m.
p-Chlorophenoxyisobutyrate	24 (5)	14 (5)	20 (8)*	13 (5)*
Clofenapate	1.6(6)	0.7 (6)	6.8 (6)	~0.6 (6)
2-(*p*-Chloromethyl)-2-(*m*-trifluoro-methylphenoxy)acetate	1.5 (5)	0.5 (5)	7.8 (6)	0.7 (3)
Halofenate	20 (3)*	20 (3)*	n.m.	n.m.
Adrenaline	1.5 (4)*	1.5 (4)*	1.5 (4)*	n.m.
D-Amphetamine	20 (6)*	10 (6)	3.2 (6)	n.m.
Fenfluramine	12 (6)	1.3 (6)	0.8 (6)	12 (5)*
Norfenfluramine	11 (5)	1.5 (5)	0.9 (6)	8 (5)*
Hydroxyethylnorfenfluramine	15 (5)	~1.3 (5)	~0.4 (6)	8 (5)*
Compound S780	2.4 (5)	0.8 (5)	0.5 (6)	n.m.
Compound S1513	1.9 (10)	~0.7 (10)	0.6 (6)	3 (6)*
Compound S1204	18 (5)*	18 (5)*	~10 (6)	n.m.
Cinchocaine	2.6 (4)	0.5 (4)	0.4 (4)	n.m.
Procaine	7 (4)*	7 (4)*	7 (4)*	n.m.
Chlorpromazine	1 (4)	0.2 (4)	0.2 (4)	~5 (6)
Demethylimipramine	1 (4)	0.3 (4)	0.3 (4)	n.m.
Mepyramine	3 (4)*	~1.9 (4)	0.7 (4)	n.m.

Note: The concentrations of compounds required to produce a 50% inhibition of the activities indicated are listed. The numbers in parentheses are the rats used. Some compounds produced no significant inhibition up to the concentrations shown and this is indicated by *; n.m. means not measured.

From Brindley, D. N. and Bowley, M., *Biochem. J.*, 148, 461, 1975. With permission.

There are two types of action that have been proposed to explain the hypotriglyceridemic actions of fenfluramine and benfluorex. These consist of direct effects in inhibiting the actions of enzymes involved in lipid metabolism and indirect effects through changing hormonal balance and thus the direction of intermediary metabolism. The direct effects that have been proposed include: (1) the inhibition of fat digestion by inhibition of the action of pancreatic lipase; (2) an inhibition of fat absorption through a decrease in the activity of monoacylglycerol acyltransferase; (3) a decrease in the rate of fatty acid synthesis; and (4) an inhibition of glycerolipid synthesis at the level of phosphatidate phosphohydrolase.[91]

The latter inhibition of phosphatidate phosphohydrolase activity is a general property of amphiphilic cationic drugs of which fenfluramine and benfluorex are members (see Section IV.D and Table 3). The extent to which the therapeutic effects of fenfluramine and benfluorex can be explained by this direct action on the phosphohydrolase is still in doubt. The major problems are first that relatively high concentrations of the drugs are required to inhibit the phosphohydrolase. Secondly, it would be predicted that the potency of the amphiphilic amines as hypotriglyceridemic agents would be governed by their hydrophobicity if they were to act solely by inhibiting the phosphohydrolase.[92] There would then be little selectivity in terms of the structure of the drugs which is not the case. It is however believed that the direct inhibition of the phosphohydrolase could be involved in the nonspecific side effects of the amphiphilic amines in producing a phospholipidosis.[91]

This condition is characterized by a generalized accumulation of phospholipids in the liposomes of a variety of tissues.[93,94] The potency of the amphiphilic cation drugs in producing these conditions is more pronounced with those compounds that are highly hydrophobic and which have long biological halflives. For example, fenfluramine and benfluorex are not potent in producing a phospholipidosis, but this condition can be seen after long term treatment with another anorectic agent, chlorphentermine.[91]

Amphiphilic cationic drugs partition into membranes by dissolving in the lipid phase and the positive charge on the amine is able to interact with the polar head group of acidic lipids. This explains why the cationic drugs interact preferentially with acidic rather than zwitterionic lipids.[95-101] This interaction can displace bivalent cations such as Ca^{2+} and Mg^{2+} from the membranes and donate a positive charge to the membrane surface. Such an interaction will interfere with the metabolism of acidic lipids. It is also likely to interfere with membrane-linked phenomena including movement, fusion, permeability, receptor function, and transport.[102,103]

The interaction of the drugs with phosphatidate inhibits the action of the phosphohydrolase, while simultaneously stimulating the activity of phosphatidate cytidylyltransferase (Sections IV.D and V). In the liver these amphiphilic cations mainly increase the accumulation of phosphatidate, whereas in other tissues increases in the production of CDP diacylglycerol, phosphatidylglycerol, and phosphatidylinositol can also be observed. Phospholipidosis is further characterized by marked increases in the accumulation of lysobisphosphatidate. This lipid is synthesized in the secondary lysosomes[104,105] by the enzymic transfer of fatty acids from phosphatidylinositol to phosphatidylglycerol or lysophosphatidylglycerol. Lysobis-phosphatidate is a characteristic marker for secondary lysosomes and its accumulation is a good indicator of the stage of development of the phospholipidosis.

Several events could contribute to the accumulation of phospholipids in lysosomes. These include the increased synthesis and accumulation of acidic phospholipids that are caused by the inhibition of phosphatidate phosphohydrolase and the stimulation of phosphatidate cytidylyltransferase.[4,91] The acidic lipids then provide additional binding sites for the amphiphilic cationic drugs. The membranes containing the lipid-drug complexes will eventually enter the lysosomal system, but it is not known whether the drugs specifically stimulate this process. The subsequent degradation of the phospholipids by various phospholipases is inhibited by the presence of the amphiphilic cationic drugs.[93,94] This may result from a combination of the abnormal phospholipid content of the membranes, the inhibitory effects of the drugs themselves, and a rise in lysosomal pH.

2. Indirect Effects

Experiments were performed to determine whether the direct effects of benfluorex in inhibiting phosphatidate phosphohydrolase could explain its observed hypolipidemic actions in vivo. In order to do this rats were treated chronically with benfluorex and then fed acutely with ethanol to stimulate the synthesis of triacylglycerol by the liver.[7] This is reflected in the increased synthesis and accumulation of triacylglycerol by the liver after ethanol feeding (Figure 6). Chronic treatment of the rats with benfluorex was also shown to decrease the conversion of phosphatidate to diacylglycerol within the liver and to increase the relative accumulation of phosphatidate.[7] This indicates that benfluorex could decrease the rate of synthesis of triacylglycerols at the level of phosphatidate phosphohydrolase.

The activity of this enzyme was also determined in the cytosolic fraction of the liver.[33] As described in Section II.B, ethanol increases the soluble phosphohydrolase activity and in these experiments increases of 4.7-fold were obtained after 6 hr. The microsomal activity was apparently only increased by only 1.4-fold[33] but since this was determined by measuring

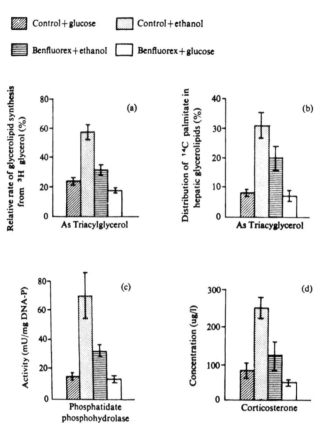

FIGURE 6.   Effects of ethanol and benfluorex on the synthesis and accumulation of hepatic triacylglycerols on the activity of hepatic phosphatidate phosphohydrolase and on the concentration of serum corticosterone (mean ± SEM values). (From Brindley, D. N. et al., *Curr. Med. Res. Opin.,* 6(Suppl. 1), 91, 1979. With permission.)

the release of inorganic phosphate the results may not reflect the true activity of the phosphohydrolase in this fraction (Chapter 1). Chronic treatment of other rats with benfluorex decreased the ethanol-induced rise in the soluble phosphohydrolase from 4.7- to 2.2-fold (Figure 6). This could not be explained by a direct inhibition of phosphatidate phosphohydrolase since the activity was measured in a cell-free system in the presence of optimum concentrations of phosphatidate.[33] It had previously been established that the inhibition by amphiphilic amines was of a competitive type with respect to phosphatidate.[92]

It therefore appeared that benfluorex was interfering with the mechanism by which ethanol increased the phosphohydrolase activity. As discussed in Section II.B, this would most likely involve the release of corticosterone in response to the ethanol load, and such an action was confirmed (Figure 6). Furthermore, there was no difference in the basal phosphohydrolase activity in the soluble fraction of rats that had been fed acutely with glucose rather than ethanol. It should be remembered that this treatment produces only a small corticosterone response that is balanced by insulin secretion.[32] The combined results in Figure 6 provide evidence that the phosphohydrolase activity is controlled by glucocorticoid release and that the increase in this activity participates in the ethanol-induced stimulation of hepatic triacylglycerol synthesis and accumulation. By contrast, benfluorex by partially preventing the stress response, diminishes the rise in the phosphohydrolase activity and also probably the release of fatty acids from adipose tissue that would stimulate triacylglycerol synthesis in the liver.

This hypothesis was investigated further by examining the effect of an acute fructose load on the responses of rats that had been maintained on a corn oil diet to magnify their stress reactions. Both benfluorex[106] and D-fenfluramine[107-108] decreased the duration of the elevated corticosterone release and the increases in the circulating concentrations of triacylglycerol in the fatty acids and glycerol that follow the feeding of fructose. Furthermore, benfluorex also decreased the fructose-induced hyperglycemia.[106] This latter event is also compatible with a decreased stress response and an increased control of metabolism by insulin.

The effects of benfluorex and D-fenfluramine in diminishing the stress response after fructose feeding appear to reside in their specific effects on the serotonergic system in the hypothalamus.[109-110] The anorectic effects of these compounds are also mediated through serotonin. Amphetamine which also produces anorexia but through the cholinergic system increases rather than decreases the duration of the fructose-induced release of corticosterone.[108] The liberation of corticosterone is thought to be initiated by the release of serotonin from the hypothalamus which in turn causes the release of corticoliberin and thereafter corticotropin. Acutely, both benfluorex and fenfluramine decrease food intake and stimulate the release of serotonin from the hypothalamus and prevent its reuptake.[109,110] This is accompanied by an acute stress response[90] and the increased release of corticosterone.[111,112] By contrast, after the chronic administration of these drugs to rats, there is no difference in food intake compared with the control and there is a decreased stress reaction. Thus, there seems to be an inverse relationship between food intake plus body weight gain and the release of stress hormones.[106-108]

The scheme outlined above seeks to explain the specific effects of fenfluramine and benfluorex as hypotriglyceridemic agents. This resides in their effects on the serotonergic system, a property that is not possessed by many other amphiphilic amines that can also inhibit directly the activity of phosphatidate phosphohydrolase. The ability of chronic treatment with fenfluramine or benfluorex to decrease the duration of the corticosterone response after feeding acute loads of ethanol or fructose is compatible with the action of benfluorex in modifying the long-term control of phosphatidate phosphohydrolase activity.[33] The diminished release of fatty acids from adipose tissue during these treatments could also decrease the acute activation of the phosphohydrolase and the recycling of the fatty acids into the triacylglycerols of very low density lipoproteins. This is discussed further in Section IV.B.

It may also be significant that the hypotriglyceridemic action of nicotinic acid depends upon its ability to decrease lipolysis in adipose tissue and the recycling of fatty acids through the liver.[113] However, the mechanism by which it does this does not depend on controlling the serotonergic response of the hypothalamus.

I. Effects of Phenobarbital

Drugs like phenobarbital are known to cause the proliferation of the smooth endoplasmic reticulum in the liver and the induction of the synthesis of certain enzymes. The injection of male rabbits for 10 days with phenobarbital produced a doubling of the microsomal protein, a 54% increase in liver weight, and an increase in the activities of aminopyrene N-demethylase, cytochrome P_{450}, and γ-glutamyltransferase.[114] Phenobarbital also increased the concentration of triacylglycerol in the liver and in the blood. These latter changes were accompanied by an increased ability of the post-mitochondrial supernatant of the liver and the microsomal fraction to synthesize glycerolipids from glycerol phosphate when expressed per liver. However, there was a decreased synthesis of glycerolipids when expressed per milligram of microsomal protein, presumably resulting from the increase in this protein concentration.

The microsomal phosphatidate phosphohydrolase activity accounted for about 85% of the total activity in the liver and the specific activity was about 20-fold higher than that of the cytosolic phosphohydrolase.[115] Both of these activities were increased in the livers of the rabbits treated with phenobarbital. The activity of the microsomal phosphohydrolase activity showed an excellent statistical correlation with the concentration of triacylglycerol in the liver and this was more significant than for the cytosolic phosphohydrolase activity. Relatively poor correlation was calculated between the activity of glycerol phosphate acyltransferase and the triacylglycerol content of the liver. These results suggest a role for the microsomal phosphohydrolase in controlling the rate of hepatic triacylglycerol synthesis although the triacylglycerol content of the liver need not necessarily parallel the rate of synthesis.[114]

Similar results were obtained when male guinea pigs were injected with phenobarbital for 7 days.[115] There were increases in the concentration of triacylglycerol in the liver and in the blood. This was accompanied by a doubling of the cytosolic phosphatidate phosphohydrolase activity, a 40% increase in the microsomal activity, and a 15% increase in the microsomal esterification of glycerol phosphate. There was a significant correlation between the microsomal phosphohydrolase activity and hepatic triacylglycerol content but not with the cytosolic activity. In the guinea pigs the microsomal phosphohydrolase accounted for about 97% of the total activity of the liver.

The injection of male rats with phenobarbital produced similar changes in the weights of the livers, the microsomal protein, and in the activities of aminopyrine N-demethylase and cytochrome P_{450}. It also increased the activity of γ-glutamyltransferase in male, but not in female rats. The hepatic concentration of triacylglycerol was also increased in starved male and female rats and in nonstarved male rats. This was again accompanied by an increase in the ability of the liver homogenates to synthesize glycerolipids from glycerol phosphate and in diacylglycerol acyltransferase activity for the male rats. There was no significant change for the male rats in the microsomal or cytosolic phosphohydrolase activities which had similar specific activities. By contrast, the microsomal phosphatidate phosphohydrolase activity of the female rats was about tenfold greater than that in the cytosolic fraction. A further discussion on the significance of these results and the differences in hepatic triacylglycerol synthesis of male and female animals is given in Section II.E.

It was concluded from the work with rats that the increase in triacylglycerol content of the liver that is produced by phenobarbital was probably related to the induction of microsomal enzymes including glycerol phosphate acyltransferase and diacylglycerol acyltransferase. The increase in hepatic triacylglycerol concentrations could than be explained by a decreased rate of secretion of very low density lipoproteins.[116] However, by contrast with the results from rabbits[114] and guinea pigs,[115] the injection of rats with phenobarbital caused a decrease in circulating triacylglycerol when the rats were starved. This decrease was not seen in fed rats.[116]

Measurements of the effects of phenobarbitol in human liver again show a high P_{450} concentration and an increase in phospholipid.[117] There was, however a decrease in the triacylglycerol concentration of the liver. By contrast, in patients with fatty livers caused by triacylglycerol or cirrhosis, there were low concentrations of phospholipid and P_{450}. Thus, there appears to be a positive correlation between the concentrations of phospholipid and P_{450}. There was an inverse correlation between P_{450} and the concentration of triacylglycerol in the liver. No significant correlation was found between the total phosphatidate phosphohydrolase activity and hepatic lipids.

It is difficult to draw clear conclusions from the work described in this section concerning how phenobarbital might modify the enzymic control of glycerolipid metabolism and whether phosphatidate phosphohydrolase might be involved. The results from rabbits[114] and guinea pigs[115] demonstrated good correlations between the activity of the phosphohydrolase in the microsomal fraction and the accumulation of hepatic triacylglycerol. However, this relationship was not observed in rats[116] and human beings.[117]

J. Effects of Some Hypolipidemic Agents Including Derivatives of Clofibrate and Phenoxypropanone

Investigations have been performed to determine whether these types of hypolipidemic agents might interfere directly with the synthesis of glycerolipids and thus decrease the subsequent secretion of very low density lipoproteins or chylomicrons. Clofibrate and some of its derivatives such as β-benzalbutyrate, clofenapate, and halofenate were able to decrease the activities of glycerol phosphate acyltransferase,[118-120] dihydroxyacetone phosphate acyltransferase,[121] and diacylglycerol acyltransferase.[120] However, clofibrate showed very little potency in inhibiting the activities of glycerol phosphate acyltransferase and diacylglycerol acyltransferase compared to its more hydrophobic derivatives, clofenapate and halofenate.[120,122] These amphiphilic anions were also much less effective in inhibiting the activity of phosphatidate phosphohydrolase compared to a series of amphiphilic cations (Table 3).

It was thought that the effects of the clofibrate type of drugs might be partly brought about by the effects of increasing the concentrations of acyl-CoA esters in microsomal membranes.[118] They could also displace fatty acids from their binding sites on albumin which in the circulation could increase the uptake of fatty acids by cells. These effects in themselves might be expected to increase rather than decrease the rate of glycerolipid synthesis since they could activate phosphatidate phosphohydrolase by promoting its translocation to microsomal membranes (Section IV.B). In fact it may be relevant in this respect that clofibrate can stimulate the synthesis of triacylglycerol in hepatocytes isolated from squirrel monkeys.[123] These authors therefore concluded that clofibrate was unlikely to exert its hypotriglyceridemic effects through an inhibition of glycerolipid synthesis.

The long-term administration of clofibrate to rats, in fact, increased the activities of the mitochondrial and microsomal glycerol phosphate acyltransferases and the peroxisomal dihydroxyacetone phosphate acyltransferase.[124] This probably relates to a general increase in the size of the hepatocytes in the drug-treated rats. However, it would be accompanied by a greater increased capacity for the oxidation of fatty acids in mitochondria and peroxisomes which might cause a smaller proportion of fatty acids to be esterified.[124]

A 5-day treatment of rats with clofibrate was shown to be effective in decreasing the incorporation of [^{14}C]glycerol into glycerolipids in the liver and intestine in vivo. These rats were also maintained on a diet rich in fructose in order to stimulate the synthesis of triacylglycerols and to produce a hypertriglyceridemia.[125] Similar effects to those observed with clofibrate in vivo were also obtained with 1,3-*bis*-(p-methylphenoxy)-2-propanone and 1-methyl-4-piperidyl *bis*-(p-chlorophenoxy) acetate. In addition, these and other related drugs were able to decrease the activities of glycerol phosphate acyltransferase and phosphatidate phosphohydrolase in cell-free preparations. There was a good correlation between the effects on the enzymes in vitro and the decreases in lipid synthesis in vivo. Hence, the direct action of these compounds in inhibiting the activities of enzymes involved in glycerolipid synthesis might be partly responsible for the hypolipidemic action. However, the authors advised caution in this interpretation since other metabolic actions are also possible.[125]

K. Effects of Thioacetoamide

Thioacetamide is a well known hepatotoxic agent that produces liver tumors. Experiments were performed to investigate the changes in lipid metabolism that might accompany this tumor development and the injury to the liver.[126] In particular, it was thought that phospholipid and membrane synthesis might be stimulated and that this might be associated with changes in polyamine synthesis (see Chapter 2, Section IV, 4). For example, the activity of ornithine decarboxylase in the liver can be increased by 50- and 40-fold when thioacetamide is administered acutely[127,128] or chronically to rats.[127] Spermine is known to form complexes with acyl-CoA esters and this may be the basis for its effect in stimulating the activity of glycerol phosphate acyltransferase and diacylglycerol acyltransferase.[129-131] In addition, spermine is able to facilitate the effects of fatty acids in promoting the activation of phosphatidate phosphohydrolase by causing its translocation to the endoplasmic reticulum (Section IV.C).

The long-term administration of thioacetamide to the rats increased the activities of phosphatidate phosphohydrolase in both the microsomal and the soluble fraction of the liver.[126] The rate of incorporation of glycerol phosphate into the total lipid fraction by the homogenates paralleled these increases in the phosphohydrolase activity. By contrast, there was no significant change in the total activity of glycerol phosphate acyltransferase after treating the rats with thioacetamide.[126]

Estimates were also made of the rate of glycerolipid synthesis in the livers in vivo by measuring the incorporation of [³H]glycerol that was injected into the portal vein of anesthetized rats.[126] There was an increase in the relative proportion of [³H]glycerol that was incorporated into triacylglycerols of the thioacetamide treated rats. However, it was unclear whether this resulted from the increased activity of phosphatidate phosphohydrolase since there was no decrease in the relative amount of [³H]glycerol that accumulated in phosphatidate. The increased rate of triacylglycerol synthesis could have been caused by increases in the availability of fatty acids and in the activity of diacylglycerol acyltransferase.[126] It is also known that the short-term administration of thioacetamide increases the accumulation of triacylglycerols in the liver. This could be a combination of an increased synthesis caused by fatty acid mobilization and a decrease in the secretion of very low density lipoproteins.[132]

Despite these relative increases in the synthesis of triacylglycerol there was no evidence for increases in phospholipid synthesis in the liver of thioacetamide-treated rats.[126] In fact, in other work, the long-term administration of thioacetamide caused a decrease in the incorporation of ³²P into phosphatidylcholine in vivo.[133]

The work with thioacetamide does not provide evidence that the increased activity of ornithine decarboxylase is necessarily associated with increases in the synthesis of membrane phospholipids.[126] Even the effects of polyamines on the activity of phosphatidate phosphohydrolase are difficult to assess. On the one hand spermine could inhibit the stimulatory effects of glucocorticoids and glucagon in the synthesis of the phosphohydrolase (Section III.A.5) and it might increase the rate of degradation of the phosphohydrolase (Section III.B). Conversely, it might acutely potentiate the effects of fatty acids in activating the phosphohydrolase by causing its association with the endoplasmic reticulum (Section IV.C).

III. LONG-TERM REGULATION OF PHOSPHATIDATE PHOSPHOHYDROLASE ACTIVITY IN VITRO

The work described in this section was designed to investigate the mechanisms that control the activity of phosphatidate phosphohydrolase. The experiments were performed in vitro under well defined conditions by using either perfused livers or isolated hepatocytes. The advantage of this approach is that the effects of hormones or other compounds can be studied in isolation. By contrast the treatment of a living animal with a particular hormone will immediately set up a series of regulatory responses including the secretion of antagonistic hormones. Thus, it is difficult to be certain of which mechanisms underlie the changes in phosphatidate phosphohydrolase activity that are reported in Section II of this chapter.

However, the disadvantage of perfused livers and isolated cells is that it is difficult to reproduce physiological conditions in vitro and to be certain that the effects being demonstrated are not artefacts to the experimental system. Fortunately, in most cases the effects of the hormones and other compounds that have been used are compatible with the changes that are found in vivo.

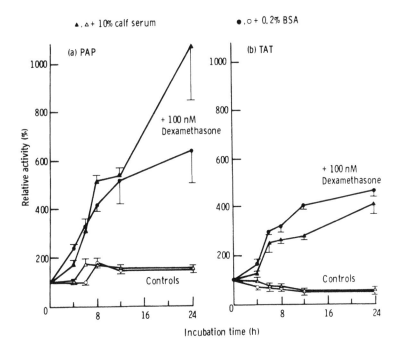

FIGURE 7.   Effects of dexamethasone on the activities of (a) phosphatidate phosphohydrolase (PAP) and (b) tyrosine aminotransferase (TAT) in rat hepatocytes. Hepatocytes
were incubated in modified Leibovitz medium containing either 10% (v/v) newborn calf
serum (▲,△) or 0.2% (w/v) fatty acid-poor bovine serum albumin (●,○) in the presence
of 100 nM dexamethasone (●,▲) or in its absence (○,△). Results are expressed as
means ± SD for three dishes from a single experiment and similar results were obtained
in a further experiment. (From Pittner, R. A. et al., *Biochem. J.*, 225, 455, 1985. With
permission.)

A. Changes in Phosphatidate Phosphohydrolase Activity that are Sensitive to Protein Synthesis Inhibitors

At present no suitable antibody against phosphatidate phosphohydrolase is available which
can be used to measure changes in the concentration of the enzyme. The experiments therefore
have to rely on the measurement of the activity of the phosphohydrolase. It is therefore not
possible to be absolutely certain that the effects that are demonstrated in this section result
directly from changes in the rate of synthesis of the phosphohydrolase. However, this is
highly likely since: (1) the effects take place over several hours; (2) they are blocked by
inhibitors of protein synthesis such as cycloheximide and actinomycin D; and (3) analogous
effects have been demonstrated with several regulatory enzymes of gluconeogenesis for
which changes in enzyme synthesis were demonstrated by immunoprecipitation.

1. Effects of Glucocorticoids

The activity of phosphatidate phosphohydrolase in perfused livers[26] and isolated hepatocytes can be increased by the presence of glucocorticoids such as cortisol, corticosterone,
and dexamethasone.[72,134-140] The increase with livers perfused with cortisol for 4 hr was 2.7-
fold.[26] This compares with increases of about twofold after 6 hr obtained with freshly prepared
hepatocytes in suspension culture.[134-137] Much larger increases of 4- to 11-fold[72,138] have
been obtained after incubating monolayer cultures of hepatocytes for 19 to 24 hr with
glucocorticoids and even after 8 hr,[138,140] these increases had reached 3.5- to 4.2-fold (Figure
7). The use of monolayer cultures of hepatocytes provides a more stable metabolic model
for these long-term studies when compared with hepatocytes in suspension culture which

are only viable for about 8 hr. The other advantage of the monolayer system is that the culture medium can be easily exchanged and the cells can be maintained in serum-free conditions.[138] This enables the investigator to use well-defined conditions and avoid interference from other hormones and compounds that might be present in the serum.

The increase in the phosphohydrolase activity that is caused by glucocorticoids probably results from a stimulation in the synthesis of the phosphohydrolase since these effects are blocked by inhibitors of protein synthesis.[72,134,138,140] This conclusion is also compatible with most of the known effects of glucocorticoids which involve the induction of protein synthesis. For example, glucocorticoids induce the synthesis of key regulatory enzymes that are involved in amino acid breakdown, urea synthesis, and gluconeogenesis. The increases in the phosphohydrolase activity are paralleled by increases in the activity of tyrosine aminotransferase.[134,138,139]

The conclusion that glucocorticoids stimulate the synthesis of phosphatidate phosphohydrolase is compatible with the increases in the activity of this enzyme that are seen in vivo: (1) in a variety of stress conditions and diabetes (Section II.A); (2) after feeding rats with ethanol and other nutrients that increase the concentrations of glucocorticoids (Section II.B); and (3) after the injection of glucocorticoids or corticotropin (Section II.C). Glucocorticoids had little[136] or no significant[137,140] effect in changing the activity of glycerol phosphate acyltransferase in cultured hepatocytes.

2. Effects of Glucagon and cAMP

A further similarity between the control of the activity of phosphatidate phosphohydrolase and tyrosine aminotransferase is seen with the effects of glucagon and cAMP. Glucagon (0.1 to 100 nM) was able to increase the activity of the phosphohydrolase which reached 1.7-fold over an 8-hr incubation period at 10 nM glucagon.[140] The effect again probably reflects changes in the rate of enzyme synthesis since it can be blocked by cycloheximide or actinomycin D. Glucagon probably acts in this respect by increasing the concentrations of cAMP with the hepatocytes since cAMP analogs when added to the incubations also increase the phosphohydrolase activity.[138,139] Secondly, the effect of glucagon is potentiated by theophylline.[140]

In the case of cAMP analogs, the effects are approximately additive to those produced by dexamethasone.[138,139] However, there was evidence of a synergistic effect[140] between the actions of glucagon and dexamethasone (Table 4). These combined effects of glucocorticoids with glucagon or cAMP analogs on the phosphohydrolase activity resemble their actions on the activities of regulatory enzymes involved in amino acid degradation and gluconeogenesis including tyrosine aminotransferase.[138,139]

Glucagon alone had no significant effect on the total activity of glycerol phosphate acyltransferase in the cultured hepatocytes but it might have slightly decreased the activity in the mitochondria and increased that in the endoplasmic reticulum.[140]

The combined effects of glucagon, cAMP, and glucocorticoids in increasing the activity of phosphatidate phosphohydrolase are compatible with the increases that are seen in this enzyme's activity in the liver in stress and diabetes (Section II.A). The smaller increases in the phosphohydrolase that can be produced by glucagon or cAMP in the absence of glucocorticoids could be responsible for the 1.7-fold increase in the phosphohydrolase activity that was seen in the livers of adrenalectomized rats that were fed with an acute intoxicating dose of ethanol.[32]

An additional long-term effect of glucagon is to increase the half-life of the phosphohydrolase activity and this will be discussed further in Section III.B. The rapid effects of glucagon and cAMP on the subcellular distribution of the phosphohydrolase are dealt with in Section IV.E.

Table 4
EFFECTS OF GLUCAGON, DEXAMETHASONE, AND INSULIN ON THE TOTAL ACTIVITY OF PHOSPHATIDATE PHOSPHOHYDROLASE (PAP) IN HEPATOCYTES AND ITS PRESENCE IN THE MEMBRANE-ASSOCIATED COMPARTMENT

| | | PAP activity in membrane-associated compartment with | | | |
| | | 0.15 mM oleate | | 0.5 mM oleate | |
Additions	(A) Total activity relative to control (%)	(B) Relative activity (%)	(C) µU/U of lactate dehydrogenase	(D) Relative activity (%)	(E) µU/U of lactate dehydrogenase
I None (control)	100 (10)	25 ± 8	98 ± 23	41 ± 4	139 ± 33
II Glucagon (10 nM)	172 ± 32 (10) I vs. II⁺·⁺	15 ± 6 I vs. II*	95 ± 52	37 ± 6	189 ± 52 I vs. II*
III Insulin (500 pM)	91 ± 13 (10)	34 ± 6 I vs. III*	1?9 ± 18	30 ± 9	102 ± 46
IV Glucagon (10 nM) + insulin (500 pM)	106 ± 11 (10) II vs. IV⁺·⁺	20 ± 7	86 ± 33	21 ± 8	108 ± 88
V Dexamethasone (100 nM)	422 ± 111 (9) I vs. V⁺·⁺	15 ± 4	273 ± 37 I vs. V*	27 ± 5	498 ± 157 I vs. V⁺
VI Dexamethasone (100 nM) + glucagon (10 nM)	830 ± 263 (9) I and II vs. VI⁺·⁺	14 ± 1	528 ± 48 I vs. VI⁺ V vs. VI*	31 ± 3	1054 ± 108 I vs. VI⁺ V vs. VI*
VII Dexamethasone (100 nM) + insulin (500 pM)	161 ± 29 (6) V vs. VII⁺⁺	17 ± 6	150 ± 28	30 ± 5	276 ± 34
VIII Dexamethasone (100 nM) + glucagon (10 nM) + insulin (500 pM)	458 ± 139 (7) VI vs. VIII⁺⁺	10 ± 3	192 ± 25 VI vs. VIII*	19 ± 1	343 ± 73 VI vs. VIII*

Note: Hepatocytes were incubated with the combinations of hormones that are indicated and after 8 hr the cells were harvested and homogenates prepared. In three independent experiments 0.15 mM or 0.5 mM oleate was added in fresh medium 15 min before the end of the incubation period. This caused the phosphohydrolase to associate with the membranes without changing the total activity. The cells were lysed with digitonin.[120] The activities in column A are expressed relative to that obtained in control incubations, which was 403 ± 152 µU/U of lactate dehydrogenase (1 U = 1 µmol/min). Columns C and E give the absolute activity in the membrane associated compartment. Columns B and D show the percentage of the total activity in this compartment; the remainder was in the cytosol. Results are expressed as means ± SD and for column A the number of independent experiments is given in parentheses. The significance of the difference between groups is indicated by *$p < 0.05$; **$p < 0.025$; ***$p < 0.01$: ⁺$p < 0.005$; ⁺⁺$p < 0.001$.

From Pittner, R. A. et al., *Biochem. J.*, 230, 525, 1985. With permission.

3. Effects of Growth Hormone

The role of growth hormone in controlling hepatic lipid metabolism has received relatively little attention. However, it can inhibit the activity of acetyl-CoA carboxylase through increasing the phosphorylation state of the enzyme[141] and it can stimulate β-oxidation.[142] Growth hormone has the ability to modify the sensitivity of tissues to insulin. Generally, it produces a state of insulin resistance whereas a deficiency of growth hormone often produced enhanced insulin sensitivity.[143-146] Although the exact mechanism for growth hormone action is not yet known, modifications of the insulin receptor[146-147] or of cAMP metabolism[148-150] have been implicated. Paradoxically, however, growth hormone also has insulin-like activity.[151-152] Increases in plasma growth hormone concentrations can produce an insulin-like effect on glucose clearance after 1 to 2 hr followed by an antagonistic effect after 6 hr.[151] These apparent contradictions might be reconciled by the observation that a growth hormone fragment (amino acids 6—13) can stimulate glycogen synthesis, whereas another fragment (amino acids 177—191) stimulates glycogenolysis.[152] Thus, variants of growth hormone may have different metabolic effects.[153]

Incubation of rat hepatocytes for 8 hr with 10 nM rat growth hormone increased the activity of phosphatidate phosphohydrolase by about 47%[154] and significant increases were obtained even when the concentration of growth hormone was decreased to 0.1 nM. These increases were not apparent until after 6 hr of incubation and they became maximum at 12 hr (Figure 8). The increases in the phosphohydrolase activity were abolished by actinomycin D and cycloheximide and so they are probably caused by a stimulation of enzyme synthesis.[154]

It therefore appears that growth hormone is acting like a "stress hormone" with respect to the control of the phosphohydrolase activity. Growth hormone secretion, as well as that of glucagon and glucocorticoids, is known to be increased in conditions such as stress, starvation, and hypoglycemia.[155] Thus, growth hormone could also contribute to the increase in phosphatidate phosphohydrolase activity that is seen in a variety of stress conditions (Section II.A). The growth hormone-induced increases in the phosphohydrolase activity are also compatible with results obtained 4 hr after injecting rats with growth hormone. In these experiments there was a 66% increase in the phosphohydrolase activity in the liver.[26]

4. Effects of Insulin

Insulin on its own had little significant effect on the long-term changes in phosphatidate phosphohydrolase activity in isolated hepatocytes.[135,136,138-140] However, it was able to antagonize the effects of glucocorticoids,[135,136,138-140] glucagon,[140] and growth hormone[154] in increasing this activity (see also Table 4). This probably occurred because it prevented the induction of the synthesis of the phosphohydrolase in a similar manner to its effects on tyrosine aminotransferase[156] and phosphoenolpyruvate carboxykinase.[157] Insulin did not reverse the increases in the phosphohydrolase activity that were produced by 8-(4-chlorophenylthio)adenosine 3′,5′-cyclic monophosphate.[138] However, this analog of cAMP was chosen because of its potency and resistance to degradation of phosphodiesterase.[138] Consequently, an action of insulin in stimulating this latter activity would not have been effective.

Insulin had no significant effects on the total activity of glycerol phosphate acyltransferase but it might have slightly decreased the activity that was found in the endoplasmic reticulum.[140]

The action of insulin on the rate of degradation of phosphatidate phosphohydrolase activity will be discussed in Section III.B and its rapid effect on the subcellular distribution of the phosphohydrolases will be dealt with in Section IV.E.

The effects of insulin in decreasing the rise in the phosphohydrolase after exposure of the hepatocytes to glucocorticoids, glucagon, and growth hormone are compatible with its action in vivo. For example, insulin injection decreases the high phosphohydrolase activity in diabetic animals (Section II.A). Secondly, the phosphohydrolase activity begins to decline in the daily cycle of the rat just after it has eaten and when the insulin concentrations are

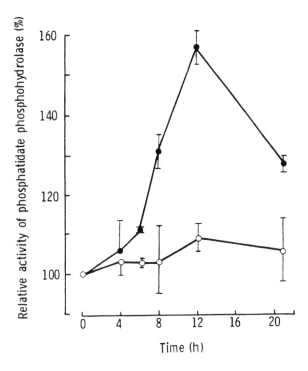

FIGURE 8.   Effect of growth hormone on the activity of phosphatidate phosphohydrolase in cultured rat hepatocytes. The cells were incubated in the absence (○) or presence (●) of 100 pM growth hormone for the times indicated. The results are the means ± ranges of the relative phosphohydrolase activity for duplicate dishes in one experiment. (Evidence for the reproducibility of the growth hormone effects is from Pittner, R. A. et al., *FEBS Lett.*, 202, 133, 1986. With permission.)

at a maximum (Section II.F). Thirdly, feeding rats acutely with fructose, sorbitol, glycerol and ethanol increases the circulating concentrations of corticosterone without significantly altering those of insulin.[32] This causes a rise in the hepatic phosphohydrolase activity. An isocaloric dose of glucose also increases the circulating concentration of corticosterone, but this is also paralleled by a rise in insulin. No significant increase is seen in the phosphohydrolase after 6 hr (Section II.B).

5. Effects of Polyamines

Polyamines have been shown to antagonize the induction of tyrosine aminotransferase that is produced by glucagon and glucocorticoids.[158] Therefore, because of the similarities in the long-term control of tyrosine aminotransferase and phosphatidate phosphohydrolase, it seemed likely that polyamines might have a similar effect on its activity.

On its own spermine had no significant effect on the phosphohydrolase activity.[159] However, 2.5 mM spermine did partially antagonize the effects of dexamethasone and glucagon in increasing the activity over a period of 8 hr. Spermidine was not effective in this respect.[159] It therefore appears that spermine has an insulin-like effect (Section III.A.4) in suppressing the stimulation in phosphatidate phosphohydrolase synthesis. A similar insulin-like action was also seen in terms of the stability of the phosphohydrolase activity in the hepatocytes where spermine accelerated the decay in the activity (Section III.B). The acute effects of polyamines in promoting the effects of fatty acids and the translocation of the cytosolic form of the phosphohydrolase to the endoplasmic reticulum will be discussed in Section IV.C.

Table 5
EFFECTS OF VARIOUS HORMONES AND COMPOUNDS ON THE HALF-LIFE OF PHOSPHATIDATE PHOSPHOHYDROLASE ACTIVITY

	Additions	Estimated half-life (hr)
I	None	6.9 ± 0.2
II	Glucagon (10 nM)	12.1 ± 2.0
III	Insulin (500 pM)	5.2 ± 0.8
		I vs. III[+]
IV	Glucagon (10 nM) + insulin (500 pM)	6.4 ± 0.5
		II vs. IV[+]
V	Dexamethasone (100 nM)	6.4 ± 0.9
VI	Dexamethasone (100 nM) + glucagon (10 nM)	9.8 ± 2.2
VII	8-Bromo-cGMP (100 μM)	8.6 ± 1.0
		I vs. VII*
VIII	Spermine (1 mM)	5.0 ± 1.1
		I vs. VIII*

Note: Hepatocytes were incubated for 12 hr with 10 nM glucagon and 100 μM corticosterone to increase phosphatidate phosphohydrolase activity. The medium was then replaced with fresh medium containing 5 μg of cycloheximide/mℓ in the presence of the compounds indicated. Results for the half-lives of the phosphohydrolase activity are means $\pm$ SD for four independent experiments. The significance of the differences between groups as calculated by using a paired t-test and it is shown by:* $p < 0.05$ and $^+$ $p < 0.02$.

From Pittner, R. A. et al., *Biochem. J.*, in press. With permission.

B. Stability of Phosphatidate Phosphohydrolase Activity

The half-life of phosphatidate phosphohydrolase activity in hepatocytes maintained in monolayer culture was determined after the inhibition of protein synthesis with cycloheximide.[160] Hepatocytes were preincubated for 12 hr with 10 nM glucagon and 100 μM corticosterone in order to increase the phosphohydrolase activity. These additions decreased the subsequent half-life of the activity in the absence of the hormones from about 13 to 7 hr. This was not accompanied by an increase in the rate of degradation of these proteins that had become labeled with [³H]leucine during the preincubation.

In the subsequent experiments all of the hepatocytes were preincubated for 12 hr with glucagon and corticosterone to produce a high phosphohydrolase activity. The addition of 10 nM glucagon or 100 μM 8-bromo-cGMP to the second incubation that contained cycloheximide decreased the rate of decline in the phosphohydrolase activity (Table 5). Dexamethasone had no significant effect. However, 500 pM insulin or 1 mM spermine increased the rate of loss of the phosphohydrolase activity. Moreover, insulin antagonized the effect of glucagon. None of these compounds significantly altered the general rate of degradation of these proteins that had been labeled with [³H]leucine during the preincubation.[160] These results demonstrate that there seems to be a fairly specific control of the stability of the phosphohydrolase by hormones.

The mechanisms by which this is brought about are not known. It is likely that the decline in the phosphohydrolase activity that occurs in the presence of cycloheximide is caused by the inactivation and the degradation of the enzyme. However, this needs to be confirmed by immunotitration when antibodies to the phosphohydrolase become available. A further complication is the fact that cycloheximide itself[161-163] and the high amino acid concentration in the incubation medium[160-162] can prevent protein degradation by autophagy. Therefore, the results concerning the phosphohydrolase activity only demonstrate part of the possible control of its stability.

However, these results do show that glucagon which stimulates increases in the phosphohydrolase activity by probably increasing its synthesis (Section III.A.2) also appears to stabilize the activity in the hepatocytes, possibly by inhibiting its degradation. Conversely, insulin (Section III.A.4) and spermine (Section III.A.5) appear to decrease both the rate of synthesis of the phosphohydrolase and its stability. These mutually antagonistic effects of insulin and glucagon could contribute to the changes in the activity of the phosphohydrolase that are observed in vivo (Section II).

It has also been shown that incubating hepatocytes for 24 hr with fatty acids increases the activity of phosphatidate phosphohydrolase. These authors suggested that this effect of fatty acids might be mediated through cGMP.[164] The action of this cyclic nucleotide in increasing the stability of the phosphohydrolase in hepatocytes could contribute to this effect.

IV. ACUTE REGULATION OF THE EXPRESSION OF PHOSPHATIDATE PHOSPHOHYDROLASE ACTIVITY IN VITRO

The work described in Section III is concerned with the regulation of phosphatidate phosphohydrolase activity probably at the level of protein synthesis and degradation. It has been shown that hormones such as the glucocorticoids, glucagon, and growth hormone are able to increase the phosphohydrolase activity. This in turn is likely to increase the potential of the liver to synthesize glycerolipids and to limit the increases in the concentrations of fatty acids and acyl-CoA esters that could occur in the liver in stress conditions. It is thought that the newly synthesized phosphohydrolase can exist in the form of a metabolically inactive reservoir in the cytosol of the liver. This form of the enzyme can be activated when it associates with the membranes of the endoplasmic reticulum.[41]

The work described in this section will review the experiments that have been performed to test this hypothesis. Furthermore, the possibility that the phosphohydrolase activity might be regulated by covalent modification will be considered.

A. Effects of Vasopressin

Vasopressin can increase the total activity of phosphatidate phosphohydrolase in suspension cultures of rat hepatocytes by about 48% in 5 min.[137] This increase appeared to depend upon the presence of Ca^{2+} in the medium since the addition of EGTA partially prevented the effects of vasopressin. The calcium ionophore, A23187, stimulated the phosphohydrolase activity by about 13%.

The increase in the phosphohydrolase activity that was produced by vasopressin was associated with increases in the rate of synthesis of phosphatidylcholine and triacylglycerol which were measured in the presence of 0.9 mM oleate.[137] The mechanisms by which vasopressin could bring about these changes are not yet known.

B. Effects of Fatty Acids, Acyl-CoA Esters, Phosphatidate, and Dicetylphosphate on the Activity and Subcellular Distribution of Phosphatidate Phosphohydrolase

The incubation of hepatocytes with fatty acids for 24 hr brings about an increase in the activity of phosphatidate phosphohydrolase[164] that may be mediated by cGMP through changes in the stability of phosphatidate phosphohydrolase (Section III.B). However, such long-term effects are unlikely to account for the increases in the phosphohydrolase activity that occur in hepatocytes that had been incubated for 60 min with 1 to 4 mM oleate.[165] The reasons for this latter effect of fatty acids are not known.

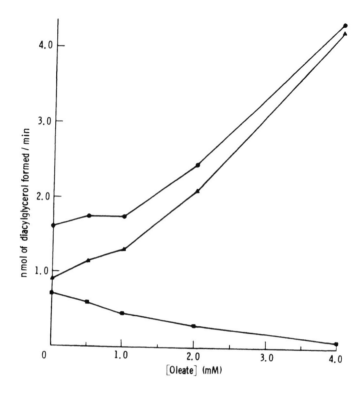

FIGURE 9.   Effect of oleate on the activity and distribution of phosphatidate phosphohydrolase in rat hepatocytes. Hepatocytes were incubated for 1 hr with the concentration of oleate that is shown and the cells were then lysed with digitonin. The total phosphohydrolase activity (●) and that which was isolated in the cytosol (■) and cell ghosts (▲) is shown. (From Cascales, C. et al., *Biochem. J.*, 219, 911, 1984. With permission.)

Perhaps the more dramatic effect of fatty acids is on the subcellular distribution of the phosphohydrolase activity within the hepatocyte. Long-chain fatty acids promote a decrease in the cytosol activity of the phosphohydrolase and an increase in the membrane-bound form of the enzyme[140,165-168] when hepatocytes are exposed to them for 15 to 60 min (Figure 9). The movement of the phosphohydrolase is paralleled by the well known increase in the synthesis of glycerolipids that is stimulated by the presence of fatty acids. This increase in substrate supply stimulates the synthesis of phosphatidate on the endoplasmic reticulum which is where the phosphohydrolase must act to produce the diacylglycerol needed for the synthesis of triacylglycerol, phosphatidylcholine, and phosphatidylethanolamine. Thus, the movement of the cytosolic form of the phosphohydrolase onto the endoplasmic reticulum is compatible with the initiation of its metabolic function.

The mechanism for this translocation of the cytosolic phosphohydrolase to the endoplasmic reticulum has been investigated with cell-free preparations from rat liver. The incubation of a mixed fraction containing microsomal membranes and soluble proteins with long-chain fatty acids or their CoA esters resulted in the association of the soluble phosphohydrolase with the membranes.[168-171] The fact that fatty acids alone, in the absence of CoA and ATP can promote the binding of the phosphohydrolase to the membranes, indicates that it is a direct effect of the acids themselves and that it need not rely on their metabolism.

Palmitate was less effective than oleate in promoting the association of the phosphohydrolase with the microsomal membranes[169] and oleate had a similar potency to linoleate, α-linolenate, arachidonate, and eicosapentenoate.[171] The CoA esters of palmitate and oleate were more effective than the corresponding acids.[169] Octanoate and octanoyl-CoA were not effective in facilitating the binding of the phosphohydrolase to the microsomal membranes.[169]

The experiments described above for this cell-free system relied simply on the collection of the membranes as a pellet after centrifugation. Separation of these membranes on a Percoll gradient showed that the phosphohydrolase had associated with membranes that contained the rotenone-insensitive cytochrome C reductase, i.e., the endoplasmic reticulum.[171] The major part of the [^{14}C]oleate that had been used to promote translocation of the phosphohydrolase was also associated with the same membrane fractions (Figure 10). This indicates that the binding of fatty acids to the membranes of the endoplasmic reticulum increases their affinity for the phosphohydrolase. This process can be reversed by incubating the membranes that contain the [^{14}C]oleate and the phosphohydrolase with increasing concentrations of fatty acid poor bovine serum albumin and then reisolating the membranes by centrifugation.[171] This results in a parallel loss of oleate and phosphohydrolase from the membranes (Figure 11).

The association of the phosphohydrolase with the membranes in the presence of oleate did not appear to depend upon the presence of bivalent cations since Ca^{2+} or Mg^{2+} (0.1 to 2.5 mM) and EGTA or EDTA (1 mM) had no significant effects.[171] Phosphate buffer increased the association of the phosphohydrolase with the microsomal membranes when they were incubated in the presence or absence of oleate. Curiously the combination of phosphate buffer with spermine in the absence or presence of oleate caused less phosphohydrolase activity to bind to the microsomal membrane.[171] The reason for this is not clear.

The addition of dicetylphosphate, which is an anionic detergent, to the combined microsomal and soluble fraction of rat liver also caused the association of the phosphohydrolase (Table 6). This indicates that the translocating effect of fatty acids is not specific but it is shared by other amphiphilic anions. These probably donate a negative charge to the microsomal membranes and thereby increase the affinity for the phosphohydrolase. A similar effect is also probably produced by the accumulation of phosphatidate on the membranes of the endoplasmic reticulum.[168] The loading of such membranes with phosphatidate also enabled the phosphohydrolase to bind to them which would be expected since this is the substrate for the enzyme. A further discussion of membrane charge and its effects on the subcellular distribution of the phosphohydrolase will take place in Section IV.D.

C. Effects of Polyamines on the Subcellular Distribution of Phosphatidate Phosphohydrolase

The addition of 0.5 to 2 mM spermine to the combined microsomal and soluble fraction of rat liver caused an increased association of the phosphohydrolase with the endoplasmic reticulum (Figure 10).[168,170,171] By contrast, 1 mM spermidine or 1 mM putrescence had no significant effect when they were added alone.[170] Spermine and to a lesser extent, spermidine also increased the effectiveness of oleate in promoting the association of the soluble phosphohydrolase with the microsomal membranes. It was concluded that spermine facilitates the interaction of the phosphohydrolase with lipophilic proteins and with the endoplasmic reticulum.[171] Previous work[172,173] with adipose tissue had demonstrated that spermine increased the retention of phosphatidate phosphohydrolase on the microsomal membranes and prevented its release into the soluble fraction (see also Chapter 3, Section VI.A).

Spermine is known to form complexes with acyl-CoA esters and this could account for the stimulation in the activities of glycerol phosphate acyltransferase[130,131] and diacylglycerol acyltransferase.[129,174] The translocation of phosphatidate phosphohydrolase could be a further factor that could account for the increased synthesis of triacylglycerols in vitro that occurs in the presence of spermine. A similar interaction could take place between fatty acids and spermine and this could be the basis for its effects on the translocation of the phosphohydrolase. In the case of acyl-CoA esters, the interaction with spermine is inhibited by ATP.[130]

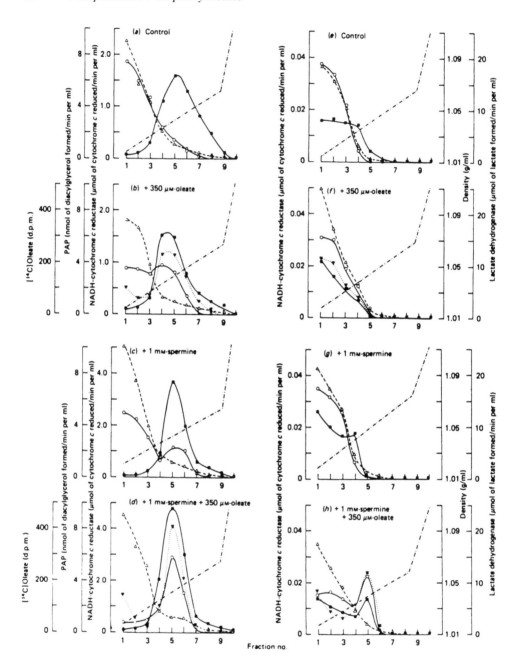

FIGURE 10.   Effects of oleate and spermine on the association of cytosolic phosphatidate phosphohydrolase with the endoplasmic reticulum of rat liver. Samples of a combined microsomal and particle-free supernatant fraction (a to d) or the particle-free supernatant fraction alone (e to h) were incubated with oleate and spermine as indicated, or in their absence (control). The fractions were then separated on Percoll gradients. The graphs show the distribution of [^{14}C]oleate (▼), density (—·—·—), and the activities of phosphatidate phosphohydrolase (○), rotenone-insensitive NADH-cytochrome c reductase (■) and lactate dehydrogenase (△). (From Hopewell, R. et al., *Biochem. J.*, 232, 485, 1985. With permission.)

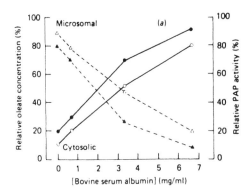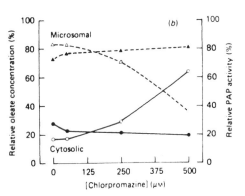

FIGURE 11. Effects of albumin and chlorpromazine on the oleate-induced translocation of phosphatidate phosphohydrolase between the cytosol and microsomal fractions of rat liver. The supernatant obtained after centrifuging homogenates from rat liver for 10 min at 18000 g was incubated for 10 min with 0.5 mM [^{14}C]oleate (12.5 mCi/mol) in order to associate the phosphohydrolase with the membranes. Albumin (a) or chlorpromazine (b) was then added and the incubations were continued for a further 10 min whereupon the cytosolic fraction was separated from the microsomal membranes by centrifugation. The relative distribution of phosphatidate phosphohydrolase activity ($\triangle$,$\bigcirc$) and [^{14}C]oleate ($\blacktriangle$,$\bullet$) in the microsomal ($\triangle$,$\blacktriangle$) and soluble fractions ($\bigcirc$,$\bullet$) is shown. (From Hopewell, R. et al., *Biochem. J.*, 232, 485, 1985. With permission.)

Table 6

EFFECTS OF OLEATE AND DICETYLPHOSPHATE ON THE TRANSLOCATION OF PHOSPHATIDATE PHOSPHOHYDROLASE BETWEEN THE CYTOSOL AND THE MICROSOMAL FRACTION[229]

Additions	Relative distribution of phosphatidate phospho-hydrolase (%)		Total activity (nmol of diacylglycerol formed/min/mg of protein)
	Soluble fraction	Microsomal fraction	
None (control)	63	37	0.83
+ Oleate (0.3 mM)	47	53	0.73
+ Oleate (1 mM)	37	63	0.83
+ Dicetylphosphate (0.3 mM)	36	64	0.76
+ Dicetylphosphate (1 mM)	28	72	0.85

Note: The supernatant obtained after centrifuging a rat liver homogenate at 18,000 × g for 10 min was incubated for 10 min at 37°C and with the compounds shown. The soluble and microsomal fractions were then separated by centrifugation and the phosphohydrolase activity was determined.[229]

The addition of ATP, GTP, CTP, AMP, and phosphate buffer also antagonized the action of spermine in promoting the association of phosphatidate phosphohydrolase with microsomal membranes.[171] However, approximately 44 mM phosphate was required to produce the same effects as 2.5 mM ATP. The addition of Mg^{2+} or Ca^{2+} at equimolar concentrations to the ATP did not reverse the effect on the spermine-induced translocation.

ATP also antagonized the effects of spermine in promoting the translocation of phosphatidate phosphohydrolase to the membranes in the presence of oleate.[171] This antagonism was not reversed by Mg^{2+} at concentrations of Mg^{2+} between 2.5 to 20 mM. It had been thought that the chelation of excess Mg^{2+} with ATP might have prevented its interaction with spermine and thereby prevented the effect on the translocation. This indicates that the affinity of spermine for ATP may be high compared to that of $Mg.^{2+}$

The effects of the nucleotides appeared to be mediated through the interaction with spermine since they have no significant effects on the translocation when they are added in the absence of polyamines.[171] Furthermore, the nucleotides did not significantly alter the distribution of the phosphohydrolase when the membranes and the soluble fraction were incubated in the presence of oleate. However, there was an inhibition by ATP of the association of the phosphohydrolase with the membranes when oleate and spermine were added together. The fact that albumin also reversed the effects of spermine alone or in the presence of oleate further suggests that the effects of spermine with oleate or other negatively charged lipids are important in promoting the association of the phosphohydrolase with the membranes.[171] This kind of specific association with negatively charged phospholipids has been described previously.[175-178] Phosphatidate could be particularly important in this respect since its accumulation in the endoplasmic reticulum would be expected to provide an ideal signal for the activation of the phosphohydrolase. However, the fact that the spermine effect on the translocation of the phosphohydrolase can probably be reversed by nucleotides and phosphatidate brings into question the extent to which this could operate in the intact cell.

At present the physiological role of polyamines in controlling glycerolipid synthesis is not well understood. Their effects in promoting the action of oleate in causing the attachment of the phosphohydrolase to the membranes of the endoplasmic reticulum could provide a stimulus for increased glycerolipid synthesis. There are also several lines of indirect evidence that suggest other mechanisms whereby polyamines might modify glycerolipid synthesis. These include effects on Ca^{2+} metabolism[179] (see also Section IV.A) and some of the similarities between the long-term control of the phosphohydrolase activity and that of ornithine decarboxylase.[126,180,181] Furthermore, a stimulation of glycerolipid synthesis would be compatible with the growth promoting effects of the polyamines.[182,183]

D. Effects of Amphiphilic Cationic Drugs on the Interaction of Phosphatidate Phosphohydrolase with the Endoplasmic Reticulum

It is well known that amphiphilic amines redirect the synthesis of glycerolipids at the level of phosphatidate metabolism.[4,41,91] This is partly brought about by the inhibition of phosphatidate phosphohydrolase and the simultaneous stimulation of phosphatidate cytidylyltransferase.[184] These events, together with other actions of the amphiphilic amines, generally decrease the rate of synthesis of triacylglycerols, phosphatidylcholine, and phosphatidylethanolamine. Conversely, the production of acidic phospholipids including phosphatidylglycerol, diphosphatidylglycerol, and phosphatidylinositol is often increased in a variety of cell types.[4,41,91]

Amphiphilic amines partition into membranes by virtue of their hydrophobic domains. They also interact particularly strongly with acidic phospholipids because of the electrostatic attraction between the positively charged nitrogen and the negative charge on the phosphate group of the acidic phospholipid.[95-101] The inhibition of phosphatidate phosphohydrolase appears to be caused by this type of interaction since (1) the potency of the amines in inhibiting the phosphohydrolase was related to their partition coefficient into the type of phosphatidate emulsion used as a substrate[92] and (2) the kinetics are compatible with the formation of a substrate-inhibitor complex.[92] It was therefore concluded that the partitioning of the amphiphilic amine into the phospholipid membrane prevented the interaction of the phosphohydrolase with its substrate.

This can now be interpreted in a slightly different way. The activation of the cytosolic form of phosphatidate phosphohydrolase involves its association with the endoplasmic reticulum in which phosphatidate is formed. It is therefore possible that amphiphilic amines

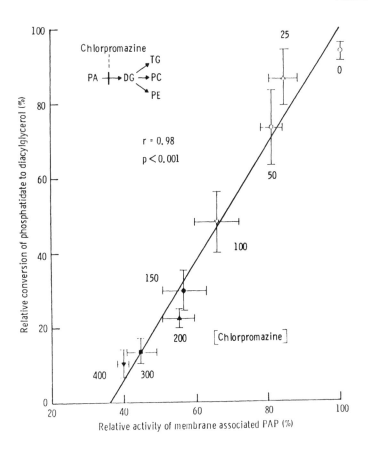

FIGURE 12. Relationship between the effects of chlorpromazine in displacing phosphatidate phosphohydrolase from the membrane-associated compartment of hepatocytes and the calculated conversion of phosphatidate to diacylglycerol. The graph shows the effects of chlorpromazine at 25 μM ($\triangle$), 50 μM ($\square$), 100 μM, ($\triangledown$), 150 μM ($\bullet$), 200 μM ($\blacktriangle$), 300 μM ($\blacksquare$) and 400 μM ($\blacktriangledown$) on the relative phosphatidate phosphohydrolase activity associated with membranes which was taken as 100% when no chlorpromazine was added ($\bigcirc$). The results are taken from three independent experiments and the values are means ± SEM. The relationship of these values is compared to the relative rate of conversion of phosphatidate (labeled with [^{3}H]glycerol) to diacylglycerol in hepatocytes that were incubated in parallel in the same experiment. (From Martin, A. et al., *Biochim. Biophys. Acta*, 876, 581, 1986. With permission.)

could prevent this activation through its association with the phosphatidate. Furthermore, the movement of the cytosolic phosphohydrolase to the endoplasmic reticulum seems to be promoted by an increase in the negative charge in the endoplasmic reticulum (Section IV.B). Therefore, the adsorption of amphiphilic amines into these membranes could oppose this change and antagonize the effects of fatty acids and acyl-CoA esters. This hypothesis was tested by using rat hepatocytes in monolayer culture and also in cell-free systems.

The addition of chlorpromazine to the cultured hepatocytes did in fact displace phosphatidate phosphohydrolase activity from the membrane compartment and it antagonized the action of oleate.[168] This action appeared to be metabolically relevant since there was a highly significant correlation between the activity of the phosphohydrolase that was membrane associated and the estimated rate of conversion of phosphatidate to diacylglycerol in the hepatocytes (Figure 12).

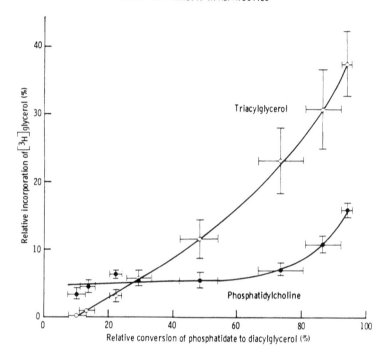

GLYCEROLIPID SYNTHESIS IN HEPATOCYTES

FIGURE 13. Relationship between the effects of chlorpromazine on the conversion of phosphatidate to diacylglycerol in hepatocytes and the synthesis of triacylglycerol and phosphatidylcholine. Chlorpromazine (0 to 400 μM) was used to decrease the calculated rate of conversion of phosphatidate (labeled with [³H]glycerol) to diacylglycerol and the consequent effect of this on the rates of synthesis of [³H]triacylglycerol (○) and [³H]phosphatidylcholine (●) is shown. The results are from five independent experiments. The values are means ± SEM and they are expressed relative to the total incorporation of [³H]glycerol into lipids which was taken as 100%. (From Martin, A. et al., *Biochim. Biophys. Acta*, 876, 581, 1986. With permission.)

The latter rate was measured in parallel incubations in which the hepatocytes were incubated under identical conditions except that the oleate and glycerol were labeled with ¹⁴C and ³H, respectively.[168] The total conversion of phosphatidate to diacylglycerol was calculated from the combined ³H in diacylglycerol, triacylglycerol, phosphatidylcholine, and phosphatidylethanolamine. Chlorpromazine (25 to 400 μM) inhibited the synthesis of phosphatidylcholine and triacylglycerol and this was accompanied by a 12-fold increase in the accumulation of labeled precursors in phosphatidate at 200 μM chlorpromazine. This represented a 76% decrease in the conversion of phosphatidate to diacylglycerol.

The relative rate of triacylglycerol synthesis at different concentrations of chlorpromazine was approximately proportional to the rate of conversion of phosphatidate to diacylglycerol (Figure 13). By contrast, phosphatidylcholine synthesis increased at higher rates of diacylglycerol formation, but it was relatively independent of this rate when this was inhibited by more than about 30% by chlorpromazine in the presence of 177 μM oleate. These results provide a further example that there is a preferential use of diacylglycerol for the synthesis of phosphatidylcholine rather than triacylglycerol when the availability of diacylglycerol is restricted. Chlorpromazine could also have inhibited phosphatidylcholine synthesis by modifying the activities of diacylglycerol acyltransferase and choline phosphotransferase.[168] However, a major site of action is the prevention of the interaction of the cytosolic phosphohydrolase with its substrate in the endoplasmic reticulum.

This was investigated further by using a mixed microsomal and soluble fraction obtained from rat liver. Chlorpromazine was shown to directly displace the phosphohydrolase and to antagonize the effects of oleate and spermine in promoting the association of the enzyme with the microsomal membranes.[168,171] Unlike albumin, chlorpromazine did not remove oleate from the membrane.[171] In fact, slightly more oleate bound to the endoplasmic reticulum in the presence of chlorpromazine. These results and the displacement of the phosphohydrolase can be understood in terms of the opposing effects of chlorpromazine and oleate on membrane charge.

Although the major effect of chlorpromazine may be brought about by changes in the charge density of the membranes it is also known to bind to calmodulin and to antagonize some of its actions. The association of calmodulin with its target enzymes involves the exposure of a lipophilic domain through a Ca^{2+}-induced conformational change.[185,186] Chlorpromazine can bind to the hydrophobic domain of calmodulin and block the activation of target enzymes by Ca^{2+}-calmodulin.[185,186] Such an action could be relevant since Ca^{2+} may be involved in controlling the activity of phosphatidate phosphohydrolase and the synthesis of triacylglycerol and phosphatidylcholine (Section IV.A). An interference with Ca^{2+}-metabolism or an interaction with phospholipid could also modify the activity of protein kinase C.[187-190] The activity of this enzyme also appears to be partly modulated by translocation to membranes and an interaction with phospholipids.[191,192] Furthermore, the product of the phosphatidate phosphohydrolase reaction is diacylglycerol which is an activator of protein kinase C. This could imply an interaction between the control of the activity of phosphatidate phosphohydrolase and protein kinase C. It has already been suggested that phosphatidate phosphohydrolase activity may be partly controlled by the action of a cAMP-dependent kinase (Section IV.E) and Ca^{2+}-dependent phosphorylations could also be important. Such effects could modify the activity and subcellular distribution of the phosphohydrolase in intact hepatocytes but they are unlikely to explain the effects of chlorpromazine in the cell-free system.

The results with chlorpromazine that are presented in this section provide further evidence that the translocation of the cytosolic phosphohydrolase to the endoplasmic reticulum is important in regulating the synthesis of glycerolipids in the liver. It also identifies a further level for which to design therapeutic agents which might be used to control this area of metabolism.

E. Effects of Glucagon, cAMP, and Insulin on the Expression of Phosphatidate Phosphohydrolase Activity

One obvious means of regulating the activity of phosphatidate phosphohydrolase is by its reversible phosphorylation. However, there is as yet no direct evidence for such a means of control. This relates to the difficulty of purifying the phosphohydrolase and in raising antibodies against it. These antibodies would enable experiments to be performed to determine whether phosphatidate phosphohydrolase can be phosphorylated under the influence of hormones.

Indirect evidence has been provided that phosphatidate phosphohydrolase activity may be modified by covalent modification by using cell-free preparations in the presence of alkaline phosphatase or ATP and Mg^{2+}. Alkaline phosphatase appeared to increase the phosphohydrolase activity whereas a presumed phosphorylation in the presence of ATP and Mg^{2+} decreased it.[193-195] Bjorkhem et al.[194] in fact preincubated samples from human liver with alkaline phosphatase in order to express what they concluded to be the maximum phosphohydrolase activity. Although the cytosolic phosphatidate phosphohydrolase activity appeared to be higher from the livers of morbidly obese patients this did not reach statistical significance. It may, however, have been more relevant to measure the activity associated with the liver membranes since this is now thought to be the functionally active form of the enzyme. Previous work with the ob/ob mouse had shown an increased activity of phosphatidate phosphohydrolase activity in liver and adipose tissue.[63,64] However, these animals have high concentrations of circulating glucocorticoids[65] which would increase the phosphohydrolase activity (Section III.A.1), whereas most human obesity is not associated with hypercortisolism. It must also be remembered that experiments with animals can be much more closely controlled than those with human beings and it is therefore easier to produce statistically significant results.

Butterwith et al.[166] were unable to confirm that preincubation of the cytosol from rat liver with alkaline phosphatase increased the phosphohydrolase activity. They made these determinations under conditions in which the alkaline phosphatase itself could not contribute to the supposed phosphatidate phosphohydrolase activity. In these experiments a preincubation in the presence of both acid or alkaline phosphatase in fact decreased the phosphatidate phosphohydrolase activity. These latter experiments suggest that phosphatidate phosphohydrolase either is composed of phosphopeptides or that some phosphate ester present in the preparations can regulate its activity.[166] For example, it has been proposed that ATP can inhibit the phosphohydrolase activity.[195] The results with alkaline phosphatase do not necessarily mean that a cAMP-dependent mechanism is involved in controlling phosphatidate phosphohydrolase activity. Furthermore, care needs to be taken when assessing the results of phosphate esters to ensure that they are not acting by complexing the Mg^{2+} that is required for phosphatidate phosphohydrolase activity.

Butterwith et al.[166] were also unable to confirm that preincubating phosphatidate phosphohydrolase with combinations of ATP, Mg^{2+}, cAMP, and cAMP-dependent protein kinase significantly altered its activity. These experiments were designed so that ATP from the preincubation should not chelate significant quantities of Mg^{2+} that were needed in the second incubation when the phosphohydrolase activity was measured. A small increase in phosphatidate phosphohydrolase activity was observed when it was preincubated with ATP. However, since a nonphosphorylating derivative of ATP gave similar increases, it was concluded that any phosphorylation of the phosphohydrolase that might occur in the presence of ATP under these conditions did not result in any obvious change in the phosphohydrolase activity.

However, this does not preclude a phosphorylation mechanism being involved in regulating the ability of the phosphohydrolase to interact with the endoplasmic reticulum. This possibility was investigated[166] by incubating hepatocytes for 60 min with a cAMP analog in the presence or absence of oleate (Table 7). Phosphatidate phosphohydrolase activity was displaced from the membrane-bound compartment by the cAMP analog, but this effect was overcome by 1 mM oleate. In fact, significantly more phosphohydrolase was associated with the membranes in the presence of the cAMP analog with 1 mM oleate than with oleate alone. These results are essentially compatible with other work in which livers were perfused with 20 μM dibutyryl cAMP and 0.5 mM oleate. This produced an increase in the phosphohydrolase activity in the microsomal fraction.[196]

The interaction of cAMP analogs and fatty acids in controlling the interaction of phosphatidate phosphohydrolase with the membrane-compartment is also reflected in the rate of glycerolipid synthesis. Thus, cAMP analogs are able to decrease the synthesis of triacylglycerol (Figure 14) and phosphatidylcholine at low fatty acid availability but these effects are reversed by adding fatty acids. These changes in the synthesis of phosphatidylcholine also involve the regulation of the ability of CTP:phosphocholine cytidylyltransferase to interact with the endoplasmic reticulum (Section V).

Experiments have also been performed to study the combination of the long-term and the short-term effects of hormones on the subcellular distribution of phosphatidate phosphohydrolase in hepatocytes. In this work the cells were incubated for 8 hr with the hormones to alter the total phosphohydrolase activity (Section III). Oleate was then added for the last

Table 7

EFFECTS OF CPT-cAMP AND OLEATE ON THE ACTIVITY AND DISTRIBUTION OF PHOSPHATIDATE PHOSPHOHYDROLASE BETWEEN THE CYTOSOL AND THE MEMBRANE-ASSOCIATED COMPARTMENT OF RAT HEPATOCYTES

Additions	Relative distribution of PAP activity (%)		Significance of difference between the relative distribution of PAP	Total PAP activity relative to (a)
	Cytosolic	Membrane-associated		
(a) None	68 ± 14	32 ± 14	(a) vs. (b), $p < 0.005$	100
			(a) vs. (c), $p < 0.05$	
			(a) vs. (d), $p < 0.02$	
(b) CPT-cAMP (0.5 mM)	86 ± 12	14 ± 12	(b) vs. (d), $p < 0.005$	112 ± 22
(c) Oleate (1 mM)	48 ± 8	52 ± 8	(c) vs. (d), $p < 0.02$	109 ± 24
(d) Oleate (1 mM) + CPT-cAMP (0.5 mM)	40 ± 9	60 ± 9		99 ± 31

Note: Hepatocytes were incubated for 1 hr at 37°C in the presence or absence of oleate and CTP-cAMP as indicated. The cells were then lysed with digitonin and the activity of phosphatidate phosphohydrolase was determined in the cytosol and membrane-associated compartment after correction for the incomplete release of lactate dehydrogenase. Results are means ± SD for five independent experiments. The significance of the differences was calculated by using a paired *t*-test. The total phosphohydrolase activity in (a) was 0.68 ± 0.57 nmol of diacylglycerol formed/min/U of lactate dehydrogenase.

From Butterwith, S. C. et al., *Biochem. J.*, 222, 487, 1984. With permission.

15 min of this incubation to promote the association of the phosphohydrolase with the membranes on which phosphatidate is synthesized.[140]

Glucagon, like the cAMP analogs, decreased the percentage of the phosphohydrolase activity that was membrane associated at the lower oleate concentration of 0.15 mM but it did not significantly alter this percentage at 0.5 mM oleate (Table 4). However, since glucagon also increases the phosphohydrolase activity in the long term, the membrane-bound activity remained fairly constant at 0.15 mM oleate and it was increased at 0.5 mM oleate. Thus, the long-term effect of glucagon (especially in the presence of glucocorticoids) is to increase the total capacity of phosphatidate phosphohydrolase and it also appears to have an acute effect that increases the concentration of fatty acid that is needed to activate the enzyme. This effect is probably mediated largely through changes in the cAMP concentration of the liver.

Insulin has the opposite effect.[140] It decreases the total phosphohydrolase activity in the long term (Sections III.A.4 and III.B), but it facilitates the interaction of the phosphohydrolase with the membranes at the lower concentration of fatty acid (Table 4). This effect could be mediated by decreasing cAMP concentrations through a stimulation of phosphodiesterase activity.[198] Alternatively, it could itself cause the phosphorylation of phosphatidate phosphohydrolase.

Such an action could be analogous to that described for the control of acetyl-CoA carboxylase which can be regulated through phosphorylation by both insulin- and cAMP-dependent mechanisms.[199] These covalent modifications change the sensitivity of the carboxylase to activation by citrate or inhibition by acyl-CoA esters. In the case of phosphatidate phosphohydrolase insulin could increase the feed-forward activation of fatty acids.

An alternative explanation for the effects of insulin and glucagon on the subcellular distribution of the phosphohydrolase is that they could modify the intracellular availability

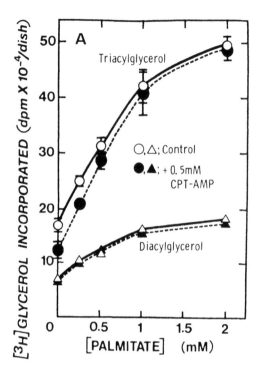

FIGURE 14. Effect of palmitate on [³H]glycerol incorporation into rat hepatocytes treated with 0.5 m*M* 8-(4-chlorophenylthio)adenosine 3′,5′-cAMP. Hepatocytes were incubated for 90 min in serum-free medium containing 100 n*M* insulin and pulsed with 10 μCi of 200 μ*M* [1,3-³H]glycerol for 60 min in the presence of 0 to 2 m*M* palmitate with (●,▲) or without (○,△) the cAMP analog. Results are means ± SEM for three dishes for triacylglycerol (○,●) or means for diacylglycerol (△,▲). (From Pelech, S. L. et al., *Biochem. J.*, 216, 129, 1983. With permission.)

of fatty acids in the hepatocyte cultures. Insulin would decrease β-oxidation and therefore make more fatty acid available to promote the translocation of the phosphohydrolase. Conversely, glucagon and cAMP could stimulate β-oxidation and decrease the net availability of fatty acids. These effects would be most obvious at low concentrations of exogenous fatty acids. However, in vivo, insulin would also decrease the transport of fatty acids to the liver from adipose tissue, whereas glucagon (together with catecholamines and corticotropin) would increase fatty acid availability.

F. Effects of Some Heat-Stable Cytosolic Proteins on the Conversion of Phosphatidate to Di- and Triacylglycerol

It has been widely observed that the addition of a cytosolic fraction to incubations that contain particulate fractions from a variety of tissues stimulates the synthesis of neutral lipids.[200] The main part of the stimulation was shown to be caused by the presence of the cytosolic phosphatidate phosphohydrolase.[10,201] Smaller stimulations were caused by what appeared to be a nonspecific protein effect.[200] There was also a heat-stable component that enhanced neutral lipid synthesis and which was shown to be caused by the presence of long-chain unsaturated fatty acids in the cytosol.[200,202] Their incorporation into phosphatidate appeared to facilitate its conversion to diacylglycerol probably because phosphatidate that contains both saturated and unsaturated acids is a better substrate than fully saturated phosphatidate.[11,202] This is discussed further in Chapter 1. However, we also know now that the association of fatty acids, particularly if unsaturated with membranes of the endoplasmic reticulum facilitates the interaction of the cytosolic phosphatidate phosphohydrolase with these membranes (Section IV.B). This might also have contributed to the stimulatory effects of the fatty acids found in the cytosol.

Later work also confirmed that cytosolic fractions from rat liver and adipose tissue stimulated the incorporation of [^{14}C]glycerol phosphate into neutral lipids.[203] Subfractionation of the cytosol showed there to be two components: one with $M_r > 150,000$ (which was probably the cytosolic phosphatidate phosphohydrolase) and a second factor that was isolated in the range expected of proteins with M_r of 8,000 to 16,000. This latter component was relatively heat stable.[203] Subsequent studies[203] showed that these heat-stable components could be inactivated by proteases and that fractions in the M_r range 35,000 to 45,000, 20,000 to 28,000, and 8,000 to 12,000 stimulated the conversion of phosphatidate to di- and triacylglycerol. The protein with M_r in the range 20,000 to 28,000 was purified 4716-fold and was shown not to have any intrinsic phosphatidate phosphohydrolase activity.

It was suggested[204] that these cytosolic proteins might promote the inter- and intracellular transfer of hydrophobic triacylglycerol precursors such as phosphatidate and diacylglycerol. The authors also claimed that the activity of the heat-stable proteins in the liver increased after refeeding rats and that it was depressed in prolonged starvation. However, there seems to be no further information about the function of these heat-stable proteins or their putative role in controlling glycerolipid synthesis.

V. THE ROLE OF ENZYME TRANSLOCATION IN COORDINATING THE REGULATION OF GLYCEROLIPID SYNTHESIS

Phosphatidylcholine synthesis is thought to be regulated in most tissues by the availability of CDP choline, the concentration of which is controlled by CTP:phosphocholine cytidylyltransferase.[205,206] This enzyme is regulated in the short-term by mechanisms that bear a remarkable similarity to those that control phosphatidate phosphohydrolase activity. The cytidylyltransferase is also thought to exist in an inactive form in the cytosol of cells and to become metabolically active when it translocates to the endoplasmic reticulum. This association with membranes is promoted by long-chain fatty acids and their CoA esters;[197,205-209] and a cAMP analog opposes these effects at low fatty acid concentrations.[205,206] It therefore appears that both phosphatidate phosphohydrolase and CTP:phosphocholine cytidylyltransferase can be classified as ambiquitous enzymes. That is they occur in different cell compartments and the control of their subcellular distribution helps to regulate metabolism.[210]

The similarity of the control of the translocation of the phosphohydrolase and the cytidylyltransferase points to a coordinated control of the synthesis of triacylglycerol and phosphatidylcholine in the liver. This can easily be understood in terms of the regulation of the production of very low density lipoproteins. These particles consist of about 60% by weight of triacylglycerol and 20% of phosphatidylcholine.

One of the main factors that controls the secretion of very low density lipoproteins is the availability of fatty acids for esterification. As the fatty acid concentration increases there is a feed-forward activation of the phosphohydrolase by the fatty acids themselves and their CoA esters. These compounds will also stimulate esterification and the subsequent accumulation of phosphatidate in the endoplasmic reticulum will further activate the phosphohydrolase and cause its association with these membranes (Section IV.B). This results in an increased production of diacylglycerol that can serve as a substrate for diacylglycerol acyltransferase and choline phosphotransferase. The cofactors for these reactions are acyl-CoA

CO-ORDINATED SYNTHESIS OF TRIACYLGLYCEROL AND PHOSPHATIDYLCHOLINE

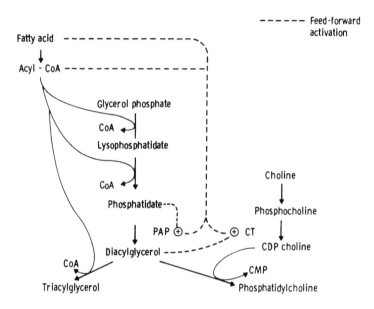

FIGURE 15.   Proposed scheme for the coordinated regulation of the synthesis of triacylglycerol and phosphatidylcholine through the control of the activations of phosphatidate phosphohydrolase and CTP:phosphocholine cytidylyltransferase by their association with the endoplasmic reticulum.

esters and CDP choline respectively. The increased availability of fatty acids and acyl-CoA esters will also stimulate the association of the cytidylyltransferase with the endoplasmic reticulum, thus increasing the production of CDP choline (Figure 15). Furthermore, the increased synthesis of diacylglycerol and its accumulation in the endoplasmic reticulum could also serve as a feed-forward signal and facilitate the activation of the cytidylyltransferase.[205,206]

The synthesis of phosphatidylcholine takes preference over that of triacylglycerol when the rate of formation of diacylglycerol is relatively low. This ensures the maintenance of membrane turnover and bile secretion which are more essential in physiological terms than the accumulation of triacylglycerol. The mechanism for the preferred synthesis of phosphatidylcholine is not certain but it could reflect a relatively low K_m of choline phosphotransferase for diacylglycerol compared with diacylglycerol acyltransferase. Triacylglycerol synthesis will be increased as the fatty acid supply rises, more diacylglycerol is produced, and acyl-CoA esters needed for the esterification become more available. As this happens it is also possible that choline may become rate limiting and the diacylglycerol will then be diverted into triacylglycerol. This is especially evident in the fatty liver that occurs in choline deficiency.

When viewed in this way, there is a competition for diacylglycerol for the synthesis of triacylglycerol and phosphatidylcholine. However, these processes can also be coordinately regulated as described above. An increased fatty acid availability causes the activation of the cytidylyltransferase which will make more CDP choline available unless there is a limitation in the supply of choline. The increased diacylglycerol that is produced by the phosphohydrolase also appears to act as a feed-forward activator for the translocation of the cytidyltransferase to the endoplasmic reticulum.[205,206] Furthermore, the diacylglycerol will also probably enhance the formation of phosphatidylcholine since it is itself a substrate.

Most workers so far have emphasized the importance of CDP choline availability in controlling phosphatidylcholine synthesis.[205,206] However, there are likely to be many situations in which choline phosphotransferase is not saturated with diacylglycerol. Some experimental evidence has been provided to support this statement.[211] The rate of phosphatidylcholine synthesis will then be proportional to the product of the concentrations of diacylglycerol and CDP choline. A simultaneous increase in the synthesis of triacylglycerol and phosphatidylcholine in the liver is likely to promote the secretion of very low density lipoproteins unless there is a defect in the supply of apolipoproteins or cholesterol for this process or in the secretory process itself.

Although the discussion so far has emphasized similarities between the control of the translocations of the phosphohydrolase and the cytidylyltransferase to the endoplasmic reticulum one might expect there to be some differences. For example, chlorpromazine is a good inhibitor for the translocation of the phosphohydrolase to the endoplasmic reticulum (Section IV.D), whereas it does not appear to be effective in this respect with the cytidylyltransferase, at least at low concentrations.[212,213] Part of the effect of chlorpromazine on the association of the phosphohydrolase with membranes appears to occur because of its selective interaction with acidic lipids such as phosphatidate. This lipid is the substrate for the phosphohydrolase and therefore should provide a high affinity binding site. Such a strict specificity for the cytidylyltransferase is unlikely although a negative charge on the membrane surface is thought to be important in enabling this enzyme to interact with membranes.[208] Chlorpromazine is a good inhibitor of the cytosolic cytidylyltransferase but its effect is antagonized by adding phospholipids to the cytosolic enzyme or when it is associated with the endoplasmic reticulum.[212,213]

As discussed in Section IV.D, chlorpromazine might also modify the rates of triacylglycerol and phosphatidylcholine synthesis by interfering with the activity of calmodulin-dependent kinases or protein kinase C. The latter enzyme activity is also probably regulated by its translocation to membranes and by its interaction with phospholipids. Diacylglycerol which is normally assumed to be produced by the stimulated breakdown of phosphoinositides, is an activator for kinase C. This kinase can also be translocated onto membranes by phorbol esters which are structurally related to diacylglycerols but which have a more prolonged effect.[191,192] It may also be significant that diacylglycerols,[205,206] phorbol esters,[214] and oleoyl alcohol[215] can stimulate the association of the cytidylyltransferase with membranes. Cornell and Vance[215,230] in fact emphasize the importance of a hydrophobic interaction of the cytidylyltransferase with membranes rather than an association with negative charges on the membrane.[209]

Phosphatidate phosphohydrolase could generate some of the diacylglycerol that might activate both protein kinase C and the cytidylyltransferase. However, theoretically it seems unlikely that diacylglycerol would be an activator of the phosphohydrolase by promoting its translocation. Normally, one would expect product inhibition rather than product activation and such an inhibition has been demonstrated.[216] Oleoyl alcohol was also found to be ineffective in translocating the phosphohydrolase.[229] However, phorbol esters have been reported to increase phosphatidate phosphohydrolase activity in adipocytes[217] but this may be through the action of protein kinase C. In turn, the phosphohydrolase could prolong the activation of kinase C by decreasing the net rate at which diacylglycerol is converted back to the phosphoinositides via phosphatidate. Thus, a complex interrelation could exist between the activation and translocation of protein kinase C, CTP:phosphocholine cytidylyltransferase, and phosphatidate phosphohydrolase.

So far, the control of glycerolipid synthesis has been discussed at the level of the balance between the rates of synthesis of triacylglycerol and phosphatidylcholine from the common diacylglycerol precursor. However, there is also an earlier branch point at the level of phosphatidate that will be immediately affected by the translocation of the cytosolic phosphohydrolase to the endoplasmic reticulum. The other routes from this branch point are the conversion of the phosphatidate to CDP-diacylglycerol or its deacylation back to glycerol phosphate. The former route is essential for the synthesis of the acidic phospholipids such as phosphatidylinositol, phosphatidylglycerol, and diphosphatidylglycerol (Chapter 1, Figure 1). The ability of cells to regulate the attachment and detachment of phosphatidate phosphohydrolase to the endoplasmic reticulum may well help to control the rate of synthesis of CDP diacylglycerol and thus the flux into acidic phospholipids. For example, a low activity of the phosphohydrolase in the endoplasmic reticulum in conditions where fatty acid availability is limited would help to maintain the essential synthesis of the acidic phospholipids.

Experiments have been performed with microsomal fractions from rat liver to investigate how changes in the concentration of phosphatidate in the membranes might alter the relative rates of synthesis of CDP diacylglycerol and diacylglycerol.[218] It was found that the rates of production of these two compounds increased in proportion to the concentration of phosphatidate and that the ratio of the rates remained fairly constant. The authors therefore concluded that both CTP:phosphatidate cytidylyltransferase and phosphatidate phosphohydrolase have a large reserve capacity. However, there are some problems with the type of experimental approach that was used. First, the ability of the cytosolic phosphohydrolase to translocate onto the endoplasmic reticulum in response to increases in the concentrations of fatty acids, acyl-CoA esters, and phosphatidate in the membranes (Section IV.B) was not known at the time. Secondly, the activity of the phosphohydrolase was determined by the release of inorganic phosphate from the membranes. The technique might not have been specific for the breakdown of phosphatidate and furthermore phospholipases of the A type can effectively degrade phosphatidate to glycerol phosphate which can then be hydrolyzed to glycerol. This can invalidate the measurement of phosphate release for the determination of phosphatidate phosphohydrolase activity in microsomal fractions of rat liver.[13]

The physiological role of this deacylation of phosphatidate in the liver[10-15] is not certain, although it could help to control the excessive accumulation of phosphatidate with the membranes of the endoplasmic reticulum. Translocation of the phosphohydrolase to these membranes would, of course, have a similar effect. The deacylase activities like that of the phosphohydrolase are found in both microsomal and soluble fractions of the liver.[12-15] It is not yet known whether the soluble deacylases are able to interact reversibly with the endoplasmic reticulum.

There is relatively little information available concerning whether the activities of CTP:phosphatidate cytidylyltransferase and phosphatidate deacylase are controlled physiologically and thereby modify the ability of phosphatidate phosphohydrolase to produce diacylglycerol. The activity of the cytidylyltransferase was not significantly changed in diabetes[19] after feeding rats acutely with ethanol[33] or chronically with diets enriched with sucrose, lard, or corn oil.[9] However, Fallon et al.[5] did detect a 25% decrease in the cytidylyltransferase activity in rats fed on a diet enriched with fructose which also increased the phosphohydrolase activity. Feeding rats acutely with glycerol, sorbitol, fructose, or ethanol had relatively little effect on the rate of phosphatidate deacylation but these nutrients did increase the phosphohydrolase activity.[14] Assuming a competition for phosphatidate, this would have stimulated the synthesis of diacylglycerol.

The most dramatic effect on the different routes of phosphatidate metabolism is produced by the family of amphiphilic cationic drugs. These displaced the phosphohydrolase from the endoplasmic reticulum and thus decreased the rate of diacylglycerol synthesis (Section IV.D). However, the drugs also stimulate the cytidylyltransferase[184] and thus the production of CDP diacylglycerol (Figure 16) and the acidic phospholipids. The amphiphilic cations are relatively less inhibitory on the activity of the phosphatidate deacylase system relative to that of the phosphohydrolase[15] and this might help to break down some of the phosphatidate that accumulates in the membranes. Amphiphilic anions (e.g., oleoyl-CoA or clofenapate) have opposite effects to amphiphilic cations in that they can stimulate the activity of phosphatidate phosphohydrolase and inhibit the cytidylyltransferase (Figure 16).

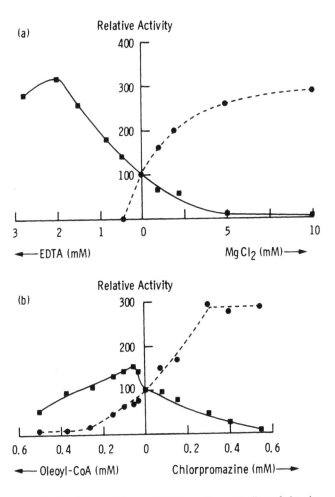

FIGURE 16. Effects of ions and EDTA on the metabolism of phosphatidate incorporated into endoplasmic reticulum membranes of rat liver. The figure shows the effects of various additions on the conversion of phosphatidate in microsomal membranes to diacylglycerol (■), or CDP-diacylglycerol (●). The amount of CDP-diacylglycerol formed in the absence of any addition was 0.48 nmol. The amount of membrane-bound phosphatidate phosphohydrolase activity was supplemented by the addition of enzyme that had been partially purified from the soluble fraction of rat liver. The amount of diacylglycerol formed in the absence of any addition ranges from 3.5 to 8.05 nmol depending upon the amount of soluble phosphohydrolase added. The membrane-bound phosphatidate already contained some Mg^{2+} which was derived from its preparation. Oleoyl-CoA was shown to interact with phosphatidate in other experiments using phosphatidate emulsions. Its partition coefficient was 24,000 ± 2300 (mean ± SD from three experiments) and the method used to determine this is described in Reference 92. (From Brindley, D. N. and Sturton, R. G., *Phospholipids: New Comp. Biochem.*, 4, 179, 1982. With permission.)

The concentration of Mg^{2+} and Ca^{2+} can also modify the direction of phosphatidate metabolism at least in vitro. Mg^{2+} stimulates both phosphatidate phosphohydrolase and phosphatidate deacylase but above the optimum concentration it can inhibit both activities.[4,15] By contrast these higher concentrations of Mg^{2+} favor the synthesis of CDP diacylglycerol[4,184] (Figure 16). Ca^{2+} can inhibit the phosphohydrolase activity whereas the deacylase system is less sensitive.[15] The production of CDP diacylglycerol is not inhibited even at high concentrations of Ca^{2+}.[219] It is not known whether these effects of Ca^{2+} and Mg^{2+} are involved in physiological regulation.

VI. INTEGRATED CONTROL OF TRIACYLGLYCEROL METABOLISM

Phosphatidate phosphohydrolase activity in the liver is increased by glucocorticoids, glucagon, and growth hormone (Section III.A). This increases the capacity for triacylglycerol synthesis especially in stress conditions (Section II.A) but this capacity need not be expressed.[72,220] It will however become metabolically active if there is a net accumulation of fatty acids and acyl-CoA esters in the liver (Section IV.B). If this accumulation were allowed to proceed unchecked it would become toxic and so the liver removes the acyl-CoA esters by producing triacylglycerols. Normally in stress conditions there is an enhanced rate of β-oxidation that also helps to limit the extent of the accumulation of fatty acids and their CoA esters. This is largely controlled by the decrease in the activity of acetyl-CoA carboxylase and the subsequent decrease in the concentration of malonyl-CoA.[28] This relieves the inhibition of carnitine palmitoyltransferase so that fatty acids can enter mitochondria for oxidation.

Severe stress and diabetes result in a large mobilization of fatty acids from adipose tissue and the liver is responsible for metabolizing a major part of this. There is a large increase in β-oxidation and the liver excretes ketone bodies. However, there is also an increase in the accumulation of fatty acids and acyl-CoA esters which cause an increased translocation of the phosphohydrolase to the endoplasmic reticulum where it is metabolically active. The acute effects of the increased cAMP concentrations mean that a higher fatty acid concentration is required to facilitate this activation. This could further help to divert fatty acids into β-oxidation. The increased triacylglycerol synthesis that will occur when net fatty acid availability increases is seen as a fatty liver in diabetes. This chapter has emphasized the role of phosphatidate phosphohydrolase in controlling triacylglycerol synthesis but there are likely to be other regulatory sites. These could include changes in the concentrations of glycerolphosphate and dihydroxyacetone phosphate which serve as substrates for the esterification and also possible changes in the activities of glycerol phosphate acyltransferase and diacylglycerol acyltransferase. High cAMP concentration may decrease the activity of the latter enzyme which could further help to decrease triacylglycerol synthesis in stress conditions.[221,222] However this effect must also be overcome by high fatty acid availability such as in ketotic diabetes in which the synthesis of triacylglycerol by the liver can be increased.[20,21] It is relevant to note in this respect that Haagsman and van Golde have shown that fatty acids stimulate the activity of diacylglycerol acyltransferase.[223] The release of glycerol by lipolysis in adipose tissue would also help to maintain or increase the concentrations of glycerol phosphate in the liver.

The triacylglycerols that are synthesized because of the increased release of fatty acids from adipose tissue in stress and diabetes can elicit and increased secretion[224] of very low density lipoproteins (Figure 17). This could be further facilitated by the action of glucocorticoids which can stimulate this process. Insulin is antagonistic in this respect.[72] The very low density lipoproteins are metabolized mainly in skeletal and cardiac muscle in starvation, diabetes, and in stress because of the relative lack of insulin action. Lipoprotein lipase activity in adipose tissue falls in these conditions, whereas that in muscle tissue is increased by the action of glucocorticoids.[226]

The function of hepatic triacylglycerol synthesis in stress conditions therefore appears to be to control the rise in fatty acid concentrations in the blood and within the liver. The liver then temporarily stores the triacylglycerol as a fatty liver or diverts them to other organs in the form of very low density lipoproteins. This process is coordinated with the secretion of glucose (from either glycogenolysis or gluconeogenesis) and of ketones from β-oxidation (Figure 17).

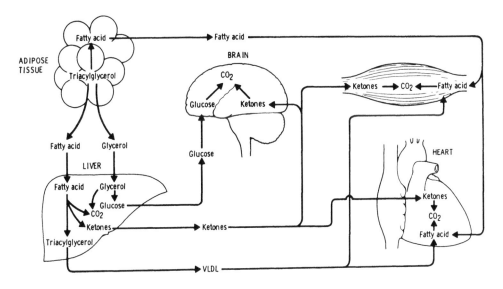

FIGURE 17.   Some effects of high concentrations of glucocorticoids and cAMP on metabolism. The solid lines represent major routes of metabolism. (From Brindley, D. N., in *Biochemistry of Lipids and Metabolism*, Vance, D. E. and Vance, J., Eds., Addison-Wesley, Reading, Mass., 1975, chap. 7. With permission.)

Ketones are used by muscle tissue and brain and spare the use of glucose which is also required by the brain and by erythrocytes. The coordinated secretion of triacylglycerols, glucose, and ketones is therefore designed to provide these tissues with supplies of energy that can be acquired in the absence of insulin. This coordination helps to explain why the increased capacity of the liver to synthesize triacylglycerols, which is expressed through phosphatidate phosphohydrolase, should bear such a striking resemblance to the control of the activity of the regulatory enzymes of gluconeogenesis.[138,140]

These events also show a link that can exist between hyperglycemia and hypertriglyceridemia. Both of these conditions can be partly caused by an increased rate of secretion of glucose or triacylglycerol from the liver accompanied by a net decrease in their removal from the blood by tissue which depends upon insulin for these processes. This can result from the lack of insulin itself, or from insulin insensitivity.

Conversely, when insulin is effective in controlling metabolism the total activity of phosphatidate phosphohydrolase in the liver is normally low. However, it can still synthesize triacylglycerols since insulin appears to enable it to translocate to the endoplasmic reticulum at the relatively low concentrations of fatty acids and their CoA esters that would occur under these conditions. Fatty acids would be derived from synthesis *de novo* and the high malonyl-CoA concentrations would suppress β-oxidation.[28] Active glycolysis would also provide the glycerol phosphate necessary for the esterification of the fatty acids. The triacylglycerols that are subsequently secreted under these conditions will be largely directed to adipose tissue where insulin will maintain a relatively high[226] lipoprotein lipase activity (Figure 18).

This chapter has attempted to review the control of phosphatidate phosphohydrolase activity in the liver and to show how this contributes to the overall regulation of fatty acid metabolism in the body. The results emphasize that large differences exist in the control of fatty acid synthesis, oxidation, and esterification. Essentially, the control of triacylglycerol synthesis is adapted to respond to changes in fatty acid availability. These fatty acids can be derived from synthesis *de novo*, from chylomicron remnants, or from adipose tissue.

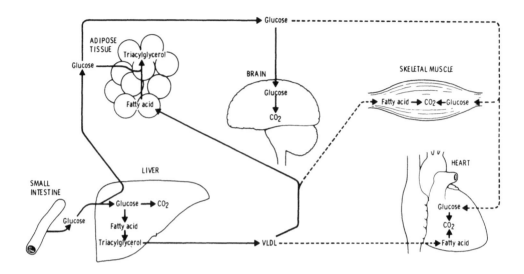

FIGURE 18.   Some effects of high concentrations of insulin on metabolism. The solid lines represent major routes of metabolism. (From Brindley, D. N., in *Biochemistry of Lipids and Metabolism*, Vance, D. E. and Vance, J., Eds., Addison-Wesley, Reading, Mass., 1975, chap. 7. With permission.)

REFERENCES

1. **Brindley, D. N.,** Intracellular phase of fat absorption, in *Biomembranes*, Vol. 4B, Smyth, D. H., Ed., Plenum Press, New York, 1974, chap. 12.
2. **Brindley, D. N.,** Digestion, absorption and transport of fats: general principles, in *Fats in Animal Nutrition*, Wiseman, J., Ed., Butterworths, London, 1984, chap. 4.
3. **Brindley, D. N.,** Regulation of fatty acid esterification in tissues, in *Control of Fatty Acid Metabolism*, Dils, R. and Knudsen, J., Eds., Pergamon Press, Oxford, 1978.
4. **Brindley, D. N. and Sturton, R. G.,** Phosphatidate metabolism and its relation to triacylglycerol biosynthesis, *Phospholipids New Comp. Biochem.*, 4, 179, 1982.
5. **Fallon, H. J., Barwick, J., Lamb, R. G., and van den Bosch, H.,** Studies on rat liver microsomal diglyceride acyltransferase and choline phosphotransferase using microsomal bound substrate: effects of high fructose intake, *J. Lipid Res.*, 16, 107, 1975.
6. **Savolainen, M. J.,** Stimulation of hepatic phosphatidate phosphohydrolase activity by a single dose of ethanol, *Biochem. Biophys. Res. Commun.*, 75, 511, 1977.
7. **Pritchard, P. H. and Brindley, D. N.,** Studies on the ethanol-induced changes in glycerolipid synthesis in rats and their partial reversal by N-(2-benzoyloxyethyl)norfenfluramine (benfluorex), *J. Pharm. Pharmacol.*, 29, 343, 1977.
8. **Glenny, H. P. and Brindley, D. N.,** The effects of cortisol, corticotropin and thyroxine on the synthesis of glycerolipids and on the phosphatidate phosphohydrolase activity in rat liver, *Biochem. J.*, 176, 777, 1978.
9. **Glenny, H. P., Bowley, M., Burditt, S. L., Cooling, J., Pritchard, P. H., Sturton, R. G., and Brindley, D. N.,** The effect of dietary carbohydrate and fat on the activities of some enzymes responsible for glycerolipid synthesis in rat liver, *Biochem. J.*, 174, 535, 1970.
10. **Smith, M. E., Sedgwick, B., Brindley, D. N., and Hübscher, G.,** The role of phosphatidate phosphohydrolase in glyceride biosynthesis, *Eur. J. Biochem.*, 3, 70, 1967.
11. **Mitchell, M. P., Brindley, D. N., and Hübscher, G.,** Properties of phosphatidate phosphohydrolase, *Eur. J. Biochem.*, 18, 214, 1971.
12. **Tzur, R. and Shapiro, B.,** Phosphatidic acid metabolism in rat liver microsomes, *Eur. J. Biochem.*, 64, 301, 1976.
13. **Sturton, R. G. and Brindley, D. N.,** Problems encountered in measuring the activity of phosphatidate phosphohydrolase, *Biochem. J.*, 171, 263, 1978.

14. **Sturton, R. G., Pritchard, P. H., Han, L.-Y., and Brindley, D. N.,** The involvement of phosphatidate phosphohydrolase and phospholipase A activities in the control of hepatic glycerolipid synthesis, *Biochem. J.,* 174, 667, 1978.

15. **Sturton, R. G. and Brindley, D. N.,** Factors controlling the metabolism of phosphatidate by phosphohydrolase and phospholipase A-type activities, *Biochim. Biophys. Acta,* 619, 494, 1980.

16. **Vavrečka, M., Mitchell, M. P., and Hübscher, G.,** The effect of starvation on the incorporation of palmitate into glycerides and phospholipids of rat liver homogenates, *Biochem. J.,* 115, 139, 1969.

17. **Mangiapane, E. H., Lloyd-Davies, K. A., and Brindley, D. N.,** A study of some enzymes of glycerolipid biosynthesis in rat liver after subtotal hepatectomy, *Biochem. J.,* 134, 103, 1973.

18. **Kinnula, V. L., Savolainen, M. J., and Hassinen, I. E.,** Hepatic triacylglycerol and fatty acid biosynthesis during hypoxia in vivo, *Acta Physiol. Scand.,* 104, 148, 1978.

19. **Whiting, P. H., Bowley, M., Sturton, R. G., Pritchard, P. H., Brindley, D. N., and Hawthorne, J. N.,** The effect of chronic diabetes, induced by streptozotocin, on the activities of some enzymes of glycerolipid synthesis in rat liver, *Biochem. J.,* 168, 147, 1977.

20. **Murthy, V. K. and Shipp, J. C.,** Synthesis and accumulation of triacylglycerol in liver of diabetic rats: effects of insulin treatment, *Diabetes,* 28, 472, 1979.

21. **Woods J. A., Knauer, T. E., and Lamb, R. G.,** The acute effects of streptozotocin-induced diabetes in rat liver glycerolipid biosynthesis, *Biochim. Biophys. Acta,* 666, 482, 1982.

22. **Murthy, V. K. and Shipp, J. C.,** Regulation of triacylglycerol synthesis in diabetic rats, *J. Clin. Invest.,* 67, 923, 1981.

23. **Lamb, R. G. and Banks, W. L.,** Effect of hydrazine exposure on hepatic triacylglycerol biosynthesis, *Biochim. Biophys. Acta,* 574, 440, 1979.

24. **Cooling, J., Burditt, S. L., and Brindley, D. N.,** Effect of treating rats with hydrazine on the circulating concentrations of corticosterone and insulin in relation to hepatic triacylglycerol synthesis, *Biochem. Soc. Trans.,* 7, 1051, 1979.

25. **Lamb, R. G. and Dewey, W. L.,** Effect of morphine exposure on mouse liver triglyceride formation, *J. Pharmacol. Exp. Ther.,* 216, 496, 1981.

26. **Lehtonen, M. A., Savolainen, M. J., and Hassinen, I. E.,** Hormonal regulation of hepatic soluble phosphatidate phosphohydrolase, *FEBS Lett.,* 99, 162, 1979.

27. **Lawson, N., Jennings, R. J., Pollard, A. D., Sturton, R. G., Ralph, S. J., Marsden, C. A., Fears, R., and Brindley, D. N.,** Effects of chronic modification of dietary fat and carbohydrate in rats. The activities of some enzymes of hepatic glycerolipid synthesis and the effects of corticotropin injection, *Biochem. J.,* 200, 265, 1981.

28. **McGarry, J. D. and Foster, D. W.,** Regulation of hepatic fatty acid oxidation and ketone body production, *Annu. Rev. Biochem.,* 49, 395, 1980.

29. **Debeer, L. J., Declercq, P. E., and Mannaerts, G. P.,** Glycerol 3-phosphate content and triacylglycerol synthesis in isolated hepatocytes from fed and starved rats, *FEBS Lett.,* 124, 31, 1981.

30. **Zammit, V. A.,** Regulation of hepatic fatty acid metabolism. The activities of mitochondrial and microsomal acyl-CoA:sn-glycerol 3-phosphate O-acyltransferase and the concentrations of malonyl:CoA, nonesterified and esterified carnitine, glycerol 3-phosphate, ketone bodies and long-chain acyl-CoA esters in livers of fed or starved pregnant, lactating the weaned rats, *Biochem. J.,* 198, 75, 1981.

31. **Savolainen, M. J. and Hassinen, I. E.,** Mechanisms for the effects of ethanol on hepatic phosphatidate phosphohydrolase, *Biochem. J.,* 176, 885, 1978.

32. **Brindley, D. N., Cooling, J., Burditt, S. L., Pritchard, P. H., Pawson, S., and Sturton, R. G.,** The involvement of glucocorticoids in regulating the activity of phosphatidate phosphohydrolase and the synthesis of triacylglycerols in the liver, *Biochem. J.,* 180, 195, 1979.

33. **Pritchard, P. H., Bowley, M., Burditt, S. L., Cooling, J., Glenny, H. P., Lawson, N., Sturton, R. G., and Brindley, D. N.,** The effects of acute ethanol feeding and of chronic benfluorex administration on the activities of some enzymes of glycerolipid synthesis in rat liver and adipose tissue, *Biochem. J.,* 166, 639, 1977.

34. **Hassinen, I. E., Savolanen, M. J., Lehtonen, M. A., Pikkukangas, A. H., and Väänänen, R. A.,** Ethanol effects on the regulation of triacylglycerol synthesis. The role of phosphatidate phosphohydrolase, in *Metabolic Effects of Alcohol,* Avogaro, P., Sirtori, C. R., and Tremoli, E., Eds., Elsevier/North-Holland, Amsterdam, 1979, 207.

35. **Savolainen, M. J. and Hassinen, I. E.,** Effect of ethanol on hepatic phosphatidate phosphohydrolase: dose-dependent enzyme induction and its abolition by adrenalectomy and pyrazole treatment, *Arch. Biochem. Biophys.,* 201, 64019, 1980.

36. **Lamb, R. G., Wood, C. K., and Fallon, H. J.,** The effect of acute and chronic ethanol intake on hepatic glycerolipid biosynthesis in the hamster, *J. Clin. Invest.,* 63, 14, 1979.

37. **Savolainen, M. J., Baraona, E., Pickkarainen, P., and Lieber, C. S.,** Hepatic triacylglycerol synthesizing activity during progression of alcoholic liver injury in the baboon, *J. Lipid Res.,* 25, 813, 1984.

38. **Wood, C. K. and Lamb, R. G.,** Effects of ethanol exposure *in vivo* and *in vitro* on glycerolipid biosynthesis of isolated or primary monolayer cultures of adult rat hepatocytes, *Fed. Proc. Fed. Am. Soc. Exp. Biol.,* 37, 420A, 1978.

39. **Pritchard, P. H., Cooling, J., Burditt, S. L., and Brindley, D. N.,** Can the alterations in serum glucocorticoid concentrations explain the effects of ethanol and benfluorex on the synthesis of hepatic triacylglycerols?, *J. Pharm. Pharmacol.,* 31, 406, 1979.

40. **Brindley, D. N., Sturton, R. G., Pritchard, P. H., Cooling, J., and Burditt, S. L.,** The mode of action of fenfluramine and its derivatives and their effects on glycerolipid metabolism, *Curr. Med. Res. Opin.,* 6(Suppl. 1), 91, 1979.

41. **Brindley, D. N.,** Intracellular translocation of phosphatidate phosphohydrolase and its possible role in the control of glycerolipid synthesis, *Prog. Lipid Res.,* 23, 115, 1984.

42. **Lamb, R. G. and Fallon, H. J.,** An enzymatic explanation for dietary induced alterations in hepatic glycerolipid metabolism, *Biochim. Biophys. Acta,* 340, 179, 1974.

43. **Yudkin, J. and Szanto, S.,** Hyperinsulinism and atherogenesis, *Br. Med. J.,* 1, 349, 1971.

44. **Bruckdorfer, K. R., Kang, S. S., and Yudkin, J.,** Plasma concentrations of insulin, corticosterone, lipids and sugars in rats fed on meals with glucose and fructose, *Proc. Nutr. Soc.,* 32, 12A, 1973.

45. **Lawson, N., Pollard, A. D., Jennings, R. J., Gurr, M. I., and Brindley, D. N.,** The activities of lipoprotein lipase and of enzymes involved in triacylglycerol synthesis in rat adipose tissue, *Biochem. J.,* 200, 285, 1981.

46. **Kako, K. J. and Peckett, S. D.,** Effect of high fat/high erucic acid diet on phosphatidate synthesis and phosphatidate phosphatase in the subcellular fractions of rat heart and liver, *Lipids,* 16, 23, 1981.

47. **Stewart, J. H. and Briggs, G. M.,** The effect of essential fatty-acid deficiency on the activity of liver phosphatidate phosphatase in rats, *Biochem. J.,* 198, 413, 1981.

48. **Blazquez, E., Castro, M., and Herrera, E.,** Effect of a high fat diet on pancreatic insulin release, glucose tolerance and hepatic gluconeogenesis in male rats, *Rev. Esp. Fisiol.,* 27, 297, 1971.

49. **Yamaguchi, K., Takashima, S., Masugama, T., and Matsuoka, A.,** Effects of the electrical stress on insulin secretion and glucose metabolism in rats fed with a high fat diet, *Endokrinol. Jpn.,* 25, 415, 1978.

50. **Lavau, M., Fried, S. R., Susini, C., and Freychet, P.,** Mechanism of insulin resistance in adipocytes of rats fed high fat diet, *J. Lipid Res.,* 20, 8, 1979.

51. **Susini, C., Lavau, M., and Herzog, J.,** Adrenaline responsiveness of glucose metabolism in insulin resistant adipose tissue of rats fed a high fat diet, *Biochem. J.,* 180, 431, 1979.

52. **Ip, C., Tepperman, H. M., Holohan, P., and Tepperman, J.,** Insulin binding and insulin response of adipocytes from rats adapted to fat feeding, *J. Lipid Res.,* 17, 588, 1976.

53. **Sun, J. V., Tepperman, H. M., and Tepperman, J.,** A comparison of insulin binding by liver plasma membranes of rats fed a high glucose or a high fat diet, *J. Lipid Res.,* 18, 533, 1977.

54. **Carroll, K. K. and Noble, R. L.,** Effects of feeding rape oil on some endocrine functions of the rat, *Endocrinology,* 51, 476, 1952.

55. **Hülsmann, W. C.,** Abnormal stress reactions after feeding diets rich in very long chain fatty acids. High levels of corticosterone and testosterone, *Mol. Cell. Endocrinol.,* 12, 1, 1978.

56. **Brindley, D. N., Cooling, J., Glenny, H. P., Burditt, S. L., and McKechnie, I. S.,** Effects of chronic modification of dietary fat and carbohydrate on the insulin, corticosterone and metabolic responses of rats fed acutely with glucose, fructose and ethanol, *Biochem. J.,* 200, 275, 1981.

57. **Macdonald, I.,** Interrelationship of dietary carbohydrates and fats on serum lipid concentrations, *Proc. Nutr. Soc.,* 30, 72A, 1971.

58. **Bruckdorfer, K. R., Kari-Kari, B. P. B., Khan, L. H., and Yudkin, J.,** Activity of lipogenic enzymes and plasma triglyceride levels in the rat and the chicken as determined by the nature of dietary fat and dietary carbohydrate, *Nutr. Metab.,* 14, 228, 1973.

59. **Lederer, J., Masri, H., and Niethals, E.,** Action lipidogène du sorbitol associé aux graisses saturées dans le régime de rat mâle, *Ann. Endocrinol. (Paris),* 39, 157, 1978.

60. **Jones, D. P. and Greene, E. A.,** Influence of dietary fat on alcoholic fatty liver, *Am. J. Clin. Nutr.,* 18, 350, 1966.

61. **Carrol, C. and Williams, L.,** Modification of ethanol induced changes in rat liver composition by the carbohydrate-fat component of the diet, *J. Nutr.,* 101, 997, 1971.

62. **Chen, N. S. C., Chen, N. C., Johnson, R. J., McGinnis, J., and Dyer, I. A.,** Effect of dietary composition on hepatic lipid accumulation of rats with chronic ethanol intake, *J. Nutr.,* 107, 1114, 1977.

63. **Jamdar, S. C., Shapiro, D., and Fallon, H. J.,** Triacylglycerol biosynthesis in the adipose tissue of the obese hyperglycaemic mouse, *Biochem. J.,* 158, 327, 1976.

64. **Fallon, H. J., Lamb, R. G., and Jamdar, S. C.,** Phosphatidate phosphohydrolase and the regulation of glycerolipid biosynthesis, *Biochem. Soc. Trans.,* 5, 37, 1977.

65. **Bray, G. A. and York, D. A.,** Hypothalamic and genetic obesity in experimental animals: an autonomic and endocrine hypothesis, *Phys. Rev.,* 59, 719, 1979.

66. **Hill, R. B. and Droke, D. W. A.,** Production of fatty liver in rats by cortisone, *Proc. Soc. Exp. Biol. Med.*, 114, 766, 1963.
67. **Ožegović, B., Rodè, B., and Milković, S.,** The role of the adrenal gland in the lipid accumulation process in the liver of rats bearing ACTH and prolactin producing tumor, *Endokrinologie*, 66, 128, 1975.
68. **Klausner, H. and Heimberg, M.,** Effect of adrenal cortisol hormones on release of triglycerides and glucose by liver, *Am. J. Physiol.*, 212, 1236, 1967.
69. **Reaven, E. S., Kolkerman, O. G., and Reaven, G. M.,** Ultrastructure and physiological evidence for corticosteroid-induced alterations in hepatic production of very low density lipoprotein particles, *J. Lipid Res.*, 15, 74, 1974.
70. **Krausz, Y., Bar-On, H., and Shafrir, E.,** Origin and pattern of glucocorticoid-induced hyperlipemia in rats. Dose-dependent bimodal changes in serum lipids and lipoproteins in relation to hepatic lipogenesis and tissue lipoprotein lipase activity, *Biochim. Biophys. Acta*, 663, 69, 1981.
71. **Cole, T. G., Wilcox, H. G., and Heimberg, M.,** Effects of adrenalectomy and dexamethasone on hepatic lipid metabolism, *J. Lipid Res.*, 23, 81, 1982.
72. **Mangiapane, E. H. and Brindley, D. N.,** Effects of dexamethansone and insulin on the synthesis of triacylglycerols and phosphatidylcholine and the secretion of very low-density lipoproteins and lysophosphatidylcholine by monolayer cultures of rat hepatocytes, *Biochem. J.*, 233, 151, 1986.
73. **Roncari, D. A. K. and Murthy, V. K.,** Effects of thyroid hormones on enzymes involved in fatty acid and glycerolipid synthesis, *J. Biol. Chem.*, 250, 4134, 1975.
74. **Young, D. L. and Lynen, F.,** Enzymatic regulation of 3-*sn*-phosphatidylcholine and triacylglycerol synthesis in states of altered lipid metabolism, *J. Biol. Chem.*, 244, 377, 1969.
75. **Soler-Argilaga, C. and Heimberg, M.,** Comparison of metabolism of free fatty acids by isolated perfused livers from male and female rats, *J. Lipid Res.*, 17, 605, 1976.
76. **Otway, S. and Robinson, D. S.,** The use of a non-ionic detergent (Triton WR 1339) to determine rates of triglyceride entry into the circulation of the rat under different physiological conditions, *J. Physiol. (London)*, 190, 321, 1967.
77. **Watkins, M. L., Fizette, N., and Heimberg, M.,** Sexual influences on hepatic secretion of triglyceride, *Biochim. Biophys. Acta*, 280, 82, 1972.
78. **Hernell, O. and Johnson, O.,** Effect of ethanol on plasma triglycerides in male and female rats, *Lipids*, 8, 503, 1973.
79. **Sidransky, H.,** Sex differences in induction of fatty liver in rat by dietary orotic acid, *Endocrinology*, 72, 709, 1963.
80. **Breen, K. J., Schenker, S., and Heimberg, M.,** Fatty liver induced by tetracycline in the rat. Dose response relationship and effect of sex, *Gastroenterology*, 69, 714, 1975.
81. **Bar-On, H., Stein, O., and Stein, Y.,** Multiple effects of cycloheximide on the metabolism of triglycerides in the liver of male and female rats, *Biochim. Biophys. Acta*, 270, 444, 1972.
82. **Savolainen, M. J., Lehtonen, M. A., Ruokonen, A., and Hassinen, I. E.,** Post-natal development and sex difference in hepatic phosphatidate phosphohydrolase activity in the rat, *Metabolism*, 30, 706, 1981.
83. **Goldberg, D. M., Roomi, M. W., and Yu, A.,** Age-related changes in hepatic triacylglycerol content and phosphatidate phosphohydrolase activities in fasting male and female rats, *Enzyme*, 30, 59, 1983.
84. **Knox, A. M., Sturton, R. G, Cooling, J., and Brindley, D. N.,** Control of hepatic triacylglycerol synthesis. Diurnal variations in hepatic phosphatidate phosphohydrolase activity and in the concentrations of circulating insulin and corticosterone in rats, *Biochem. J.*, 180, 441, 1979.
85. **Jamdar, S. C., Moon, M., Bow, S., and Fallon, H. J.,** Hepatic lipid metabolism. Age-related changes in triglyceride metabolism, *J. Lipid Res.*, 19, 763, 1978.
86. **Coleman, R. A. and Haynes, E. B.,** Selective changes in microsomal enzymes of triacylglycerol and phosphatidylcholine synthesis in fetal and postnatal rat liver, *J. Biol. Chem.*, 258, 450, 1983.
87. **Coleman, R. A. and Haynes, E. B.,** Hepatic monoacylglycerol acyltransferase. Characterization of an activity associated with the suckling period in rats, *J. Biol. Chem.*, 259, 8934, 1984.
88. **Coleman, R. A. and Haynes, E. B.,** Subcellular location and topography of rat hepatic monoacylglycerol acyltransferase activity, *Biochim. Biophys. Acta*, 834, 180, 1985.
89. **Coleman, R. A. and Haynes, E. B.,** Monoacylglycerol acyltransferase. Evidence that the activities from rat intestine and suckling liver are tissue-specific isoenzymes, *J. Biol. Chem.*, 261, 224, 1986.
90. **Dannenburg, W. N.,** Metabolic effects of phenylethylamine drugs on glucose and fatty acids in various tissues, in *Biochemical Pharmacology of Obesity*, Curtis-Prior, P. B., Ed., Elsevier, Amsterdam, 1983, chap. 11.
91. **Brindley, D. N.,** Phenylethylamines and their effects on the synthesis of fatty acids, triacylglycerol and phospholipids, in *Biochemical Pharmacology of Obesity*, Curtis-Prior, P. B., Ed., Elsevier, Amsterdam, 1983, chap. 12.
92. **Bowley, M., Cooling, J., Burditt, S. L., and Brindley, D. N.,** The effects of amphiphilic cationic drugs and inorganic ions on the activity of phosphatidate phosphohydrolase, *Biochem. J.*, 165, 447, 1977.

93. **Lüllmann, H., Lüllmann-Rauch, R., and Wasserman, O.,** Drug-induced phospholipidosis, *Crit. Rev. Toxicol.,* 4, 185, 1979.

94. **Lüllman, H., Lüllmann-Rauch, R., and Wasserman, O.,** *Biochem. Pharmacol.,* 27, 1103, 1978.

95. **Giotta, G. J., Chan, D. S., and Wang, H. H.,** Binding of spin-labelled local anaesthetics to phosphatidylcholine and phosphatidylserine liposomes, *Arch. Biochem. Biophys.,* 163, 453, 1974.

96. **Hauser, H., Penkett, S. A., and Chapman, D.,** Nuclear magnetic resonance spectroscopy studies of procaine hydrochloride and tetracaine hydrochloride at lipid-water interfaces, *Biochim. Biophys. Acta,* 183, 466, 1969.

97. **Lee, A. G.,** Interactions between anaesthetics and lipid mixture amines, *Biochim. Biophys. Acta,* 448, 34, 1976.

98. **Paphadjopoulos, D., Jacobson, K., Poste, G., and Shepherd, G.,** Effect of local anaesthetics on membrane properties. I. Changes in fluidity of phospholipid bilayers, *Biochim. Biophys. Acta,* 394, 504, 1975.

99. **Ragazzi, M., Gorio, A., and Peluchetti, D.,** Influence of Ca^{2+} on interaction of anaesthetic drugs with artificial phospholipid membranes, *Experientia,* 31, 567, 1975.

100. **Carey, M. C., Hiron, P. C., and Small, D. M.,** A study of the physicochemical interactions between biliary lipids and chlorpromazine hydrochloride. Bile salt precipitation as a mechanism of phenothiazine-induced bile secretion failure, *Biochem. J.,* 153, 519, 1976.

101. **Lüllmann, H. and Wehling, M.,** Binding of drugs to different polar lipids in vitro, *Biochem. Pharmacol.,* 28, 3409, 1979.

102. **Brindley, D. N., Allan, D., and Michell, R. H.,** The redirection of glyceride and phospholipid synthesis by drugs including chlorpromazine, fenfluramine, imipramine, mepyramine and local anaesthetics, *J. Pharm. Pharmacol.,* 27, 462, 1975.

103. **Brindley, D. N., Bowley, M., Sturton, R. G., Pritchard, P. H., Burditt, S. L., and Cooling, J.,** Mode of action of fenfluramine and derivatives and their effects on glycerolipid metabolism, in *Central Mechanisms of Anorectic Drugs,* Garattini, S. and Samanin, R., Eds., Raven Press, New York, 1978, 301.

104. **Matsuzawa, Y., Poorthuis, B. J. H. M., and Hostetler, K. Y.,** Mechanism of the phosphatidylinositol stimulation of lysosomal bis(monoacylglyceryl)phosphate synthesis, *J. Biol. Chem.,* 253, 6655, 1978.

105. **Matsuzawa, Y. and Hostetler, K. J.,** Inhibition of lysosomal phospholipase A and phospholipase C by chloroquine and 4,4'-bis(diethylaminoethoxy), α, β-diethylphenylethane, *J. Biol. Chem.,* 255, 5190, 1980.

106. **Brindley, D. N., Cousins, C., Finnerty, C., Saxton, J., and Mangiapane, E. H.,** Chronic treatment of rats with benfluorex decreases stress responses and has a hypotriglyceridaemic and hypoglycaemic effect, *G. Arterioscler.,* Suppl. 1, 32, 1985.

107. **Brindley, D. N., Saxton, J., Shahidullah, H., and Armstrong, M.,** Possible relationships between changes in body weight set-point and stress metabolism after treating rats chronically with D-fenfluramine. Effects of feeding rats acutely with fructose on the metabolism of corticosterone, glucose, fatty acids, glycerol and triacylglycerol, *Biochem. Pharmacol.,* 34, 1265, 1985.

108. **Brindley, D. N., Saxton, J., Shahidullah, H., Armstrong, M., and Mangiapane, E. H.,** Dextrofenfluramine: relationship between decreases of the body weight set-point and metabolic effects, in *Metabolic Complications of Human Obesities,* Vague, J., Björntorp, P., Guy-Grand, B., Rebuffe-Scrive, M., and Vague, P., Eds., Excerpta Medica Int. Cong. Series 682, Excerpta Medica, Amsterdam, 207.

109. **Borsini, F. Bendotti, C., Aleotti, A., Samanin, R., and Garattini, S.,** D-fenfluramine and D-norfenfluramine reduce food intake by acting on different serotonin mechanisms in rat brain, *Pharmacol. Res. Commun.,* 14, 671, 1982.

110. **Samanin, R.,** Drugs affecting serotonin and feeding, in *Biochemical Pharmacology of Obesity,* Curtis-Prior, P. B., Ed., Elsevier, Amsterdam, 1983, chap. 14.

111. **Shettini, G., Quattrone, A., Di Renzo, G. F., and Prezosi, P.,** Effects of selective degeneration of brain serotonin-containing neurones on plasma corticosterone levels: studies with d-fenfluramine, *Pharmacol. Res. Commun.,* 11, 545, 1979.

112. **Fuller, R. W., Snoddy, H. D., and Clemens, J. A.,** Elevation by fenfluramine of 3,4-dihydroxyphenylacetic acid in brain and of corticosterone and prolactin in serum of fenfluramine pre-treated rats, *Pharmacol. Res. Commun.,* 13, 275, 1981.

113. **Fears, R.,** Drug treatment of hyperlipidaemia, *Drugs Today,* 20, 257, 1984.

114. **Goldberg, D. M., Roomi, M. W., Yu, A., and Roncari, D. A. K.,** Effects of phenobarbital upon triacylglycerol metabolism in the rabbit, *Biochem. J.,* 192, 165, 1980.

115. **Goldberg, D. M., Yu, A., Roomi, M. W., and Roncari, D. A. K.,** Effects of phenobarbital upon triacylglycerol metabolism in the guinea pig, *Can. J. Biochem.,* 59, 48, 1981.

116. **Goldberg, D. M., Roomi, M. W., Yu, A., and Roncari, D. A. K.,** Triacylglycerol metabolism in the phenobarbital-treated rat, *Biochem. J.,* 196, 337, 1981.

117. **Savolainen, M. J., Arranto, A. A., Hassinen, I. E., Luoma, P. V., Pelkonen, R. O., and Sotaniemmi, E. A.,** Relationship between lipid composition and drug metabolising capacity of human liver, *Eur. J. Clin. Pharmacol.,* 27, 727, 1985.

118. **Lamb, R. G. and Fallon, H. J.,** Inhibition of monoacylglycerophosphate formation by chlorphenoxyi-sobutyrate and β-benzalbutyrate, *J. Biol. Chem.,* 247, 1281, 1972.

119. **Fallon, H. J., Adams, L. L., and Lamb, R. G.,** A review of studies on the mode of action of clofibrate and betabenzalbutyrate, *Lipids,* 7, 106, 1972.

120. **Brindley, D. N. and Bowley, M.,** Drugs affecting the synthesis of glycerides and phospholipids in rat liver, *Biochem. J.,* 148, 461, 1975.

121. **Bowley, M. and Brindley, D. N.,** Selective inhibition by clofenapate of glycerolipid synthesis via the esterification of dihydroxyacetoone phosphate in rat liver slices, *Int. J. Biochem.,* 7, 141, 1976.

122. **Horney, C. J. and Margolis, S.,** Comparison of the effects of clofibrate and halofenate (MK-185) in isolated rat hepatocyte, *Atherosclerosis,* 19, 381, 1974.

123. **Capuzzi, D. M., Lackman, R. D., Uberti, M. O., and Reed, M. A.,** *Biochem. Biophys. Res. Commun.,* 60, 149, 1974.

124. **Pollard, A. D. and Brindley, D. N.,** Effect of chronic clofibrate feeding on the activities of enzymes involved in glycerolipid synthesis and in peroxisomal metabolism in the rat, *Biochem. Pharmacol.,* 31, 1650, 1982.

125. **Lamb, R. G., Wyrick, S. D., and Piantadosi, C.,** Hypolipidemic activity of in vitro inhibitors of hepatic and intestinal sn-glycerol 3-phosphate acyltransferase and phosphatidate phosphohydrolase, *Atherosclerosis,* 27, 147, 1977.

126. **Cascales, C., Martin-Sanz, P., Pittner, R. A., Hopewell, R., Brindley, D. N., and Cascales, M.,** Effects of an antitumoural rhodium complex on thioacetamide-induced liver cancer in rats. Changes in the activities of ornithine decarboxylase, tyrosine, aminotransferase and of enzymes involved in fatty acid and glycerolipid synthesis, *Biochem. Pharmacol.,* 35, 2655, 1986.

127. **Pegg, A. E., Conover, C., and Wrona, A.,** Effects of aliphatic diamines on rat liver ornithine decarboxylase activity, *Biochem. J.,* 160, 651, 1978.

128. **Seeley, J. E. and Pess, A. E.,** Effect of 1,3-diaminopropane on ornithine decarboxylase enzyme protein in thioacetamide-treated rat liver, *Biochem. J.,* 216, 701, 1984.

129. **Jamdar, S. C.,** Glycerolipid biosynthesis in rat adipose tissue. Effect of polyamines on triglyceride synthesis, *Arch. Biochem. Biophys.,* 182, 723, 1977.

130. **Jamdar, S. C.,** Hepatic lipid metabolism: effect of spermine, albumin, and Z-protein on microsomal lipid formation, *Arch. Biochem. Biophys.,* 195, 81, 1979.

131. **Bates, E. J. and Saggerson, E. D.,** Effects of spermine and albumin on hepatic mitochondrial and microsomal glycerol phosphate acyltransferase activity, *Biochem. Soc. Trans.,* 9, 57, 1981.

132. **Franke, H., Zimmermann, T., and Dargel, R.,** Qualitative and quantitative changes in hepatic lipoprotein particles following acute injury of the rat liver induced by thioacetamiide, *Virchow's Arch. B,* 44, 99, 1983.

133. **Palacios, E., Osada, J., Arce, C., Aylagas, H., and Santos-Ruiz, A.,** Effecto de la tioacetamida sobre la incorpación de ^{32}P a fosfolipidos de higado de rata, *Rev. Esp. Fisiol.,* 35(Suppl.), 135, 1982.

134. **Jennings, R. J., Lawson, N., Fears, R., and Brindley, D. N.,** Stimulation of the activities of phosphatidate phosphohydrolase and tyrosine aminotransferase in rat hepatocytes by glucocorticoids, *FEBS Lett.,* 133, 119, 1981.

135. **Lawson, N., Jennings, R. J., Fears, R., and Brindley, D. N.,** Antagonistic effects of insulin on the corticosterone-induced increase of phosphatidate phosphohydrolase activity in isolated rat hepatocytes, *FEBS Lett.,* 143, 9, 1982.

136. **Lawson, N., Pollard, A. D., Jennings, R. J., and Brindley, D. N.,** Effects of corticosterone and insulin on enzymes of triacylglycerol synthesis in isolated rate hepatocytes, *FEBS Lett.,* 146, 204, 1982.

137. **Pollard, A. D. and Brindley, D. N.,** Effects of vasopressin and corticosterone on fatty acid metabolism and on the activities of glycerol phosphate acyltransferase and phosphatidate phosphohydrolase in rat hepatocytes, *Biochem. J.,* 217, 461, 1984.

138. **Pittner, R. A., Fears, R., and Brindley, D. N.,** Effects of cyclic AMP, glucocorticoids and insulin on the activities of phosphatidate phosphohydrolase, tyrosine aminotransferase and glycerolkinase in isolated rat hepatocytes in relation to the control of triacylglycerol synthesis and gluconeogenesis, *Biochem. J.,* 225, 455, 1985.

139. **Pittner, R. A., Mangiapane, E. H., Fears, R., and Brindley, D. N.,** Control of the activities of phosphatidate phosphohydrolase and tyrosine aminotransferase by gluocorticoids, cyclic AMP and insulin in rat hepatocytes, *Biochem. Soc. Trans.,* 13, 159, 1985.

140. **Pittner, R. A., Fears, R., and Brindley, D. N.,** Interactions of insulin, glucagon and dexamethansone in controlling the activity of glycerol phosphate acyltransferase and the activity and subcellular distribution of phosphatidate phosphohydrolase in cultured rat hepatocytes, *Biochem. J.,* 230, 525, 1985.

141. **Bornstein, J., Ng, F. M., Heng, D., and Wang, K. P.,** Metabolic actions of pituitary growth hormone 1. Inhibition of acetyl-CoA carboxylase by human growth hormone and a carboxyl terminal part sequence acting through a second messenger, *Acta Endocrinol.,* 103, 479, 1983.

142. **Clejan, S.,** Effects of growth hormone on fatty acid oxidation, *Biophys. J.,* 47, A199, 1985.

143. **Parkes, M. J. and Bassett, J. M.,** Antagonism by growth hormone of insulin action in fetal sheep, *J. Endocrinol.,* 105, 379, 1985.

144. **Ruzza, R. A., Mandanno, L. J., and Gernick, J. E.,** Effects of growth hormone on insulin action in man. Mechanisms of insulin resistance, impaired suppression of glucose production and impaired stimulation of glucose utilization, *Diabetes,* 31, 663, 1982.

145. **Bratusch-Marrain, P. R., Smith, D., and DeFronzo, R. A.,** The effect of growth hormone on glucose metabolism and insulin secretion in man, *J. Clin. Endocrinol. Metab.,* 55, 973, 1982.

146. **Kahn, C., Goldfine, I., Neville, D., and De Meyts, P.,** Alterations in insulin binding induced by changes *in vivo* in the levels of glucocorticoids and growth hormone, *Endocrinology,* 103, 1054, 1978.

147. **Olefsky, J. M., Johnson, J., Lui, F., Jen, P., and Reaven, G. M.,** The effect of acute and chronic dexamethasone administration on insulin binding to isolated rat hepatocytes and adipocytes, *Metabolism,* 24, 517, 1975.

148. **Albertson-Wikland, K. and Roseberg, S.,** Inhibition of adenylate cyclase activity in muscles by growth hormone, *Endocrinology,* 111, 1855, 1982.

149. **Leppert, P., Guillory, J., Russo, L. J., and Moore, W. V.,** In vivo effect of human growth hormone on hepatic adenylate cyclase activity, *Endocrinology,* 109, 990, 1981.

150. **Byus, V. C., Haddox, M. R., and Russel, D. H.,** Activation of cyclic AMP-dependent protein kinase(s) by growth hormone in liver and adrenal gland of rat, *J. Cyclic Nucleotide Res.,* 4, 45, 1978.

151. **MacGorman, L. R., Rizza, R. A., and Gerich, J. E.,** Physiological concentrations of growth hormone exert insulin-like and insulin antagonistic effects on both hepatic and extrahepatic tissues, *J. Clin. Endocrinol. Metab.,* 53, 556, 1981.

152. **Newman, J. I., Armstrong, J. McD., and Bornstein, J.,** Effects of part sequences of human growth hormone on in vivo hepatic glycogen metabolism in the rat, *Biochim. Biophys. Acta,* 544, 234, 1978.

153. **Lewis, E. J., Calle, P., and Wicks, W. D.,** Differences in rates of tyrosine aminotransferase deinduction with cyclic AMP and glucocorticoids, *Proc. Natl. Acad. Sci. U.S.A.,* 79, 5778, 1982.

154. **Pittner, R. A., Bracken, P., Fears, R., and Brindley, D. N.,** Insulin antagonises the growth hormone mediated increase in the activity of phosphatidate phosphohydrolase in isolated rat hepatocytes, *FEBS Lett.,* 202, 133, 1986.

155. **Reichlin, A.,** *Handbook of Physiology,* Vol. 2, Knobit, K. and Sawyer, W. H., Eds., Am. Physiol. Soc., Washington, D.C., 1974, 407.

156. **Spencer, C. J., Heaton, J. H., Gelehrter, T. A., Richardson, K. I., and Garwin, J. L.,** Insulin selectively slows the degradation rate of tyrosine aminotransferase, *J. Biol. Chem.,* 253, 7677, 1978.

157. **Granner, D., Andreone, T., Sasaki, K., and Beale, E.,** Inhibition of transcription of the phosphoenolpyruvate carboxykinase gene by insulin, *Nature (London),* 308, 549, 1983.

158. **Auberger, P., Samson, M., and Le Cam, A.,** Inhibition of hormonal induction of tyrosine aminotransferase by polyamines in freshly prepared rat hepatocytes, *Biochem. J.,* 214, 679, 1983.

159. **Pittner, R. A., Bracken, P., Fears, R., and Brindley, D. N.,** Spermine antagonises the effects of dexamethasone and glucagon in increasing the activity of phosphatidate phosphohydrolase in isolated hepatocytes, *FEBS Lett.,* 207, 42, 1986.

160. **Pittner, R. A., Fears, R., and Brindley, D. N.,** Effects of insulin, glucagon, dexamethasone, cyclic GMP and spermine on the stability of phosphatidate phosphohydrolase activity in cultured rat hepatocytes, *Biochem. J.,* 240, 253, 1986.

161. **Woodside, K. H.,** Effects of cycloheximide on protein degradation and gluconeogenesis in the perfused rat liver, *Biochim. Biophys. Acta,* 421, 70, 1976.

162. **Seglen, P. O.,** Protein-catabolic state of isolated rat hepatocytes, *Biochim. Biophys. Acta,* 496, 182, 1977.

163. **Brookes, C., Wright, A., Evans, P. J., and Mayer, R. J.,** Aflatoxin B1 — effect on the synthesis and degradation of mitochondrial proteins in hepatocyte monolayers, *Carcinogenesis,* 5, 759, 1984.

164. **Lamb, R. G. and McCue, S. B.,** The effect of fatty acid exposure on the biosynthesis of glycerolipids by cultured hepatocytes, *Biochim. Biophys. Acta,* 753, 356, 1983.

165. **Cascales, C., Mangiapane, E. H., and Brindley, D. N.,** Oleic acid promotes the activation and translocation of phosphatidate phosphohydrolase from the cytosol to particulate fractions of isolated rat hepatocytes, *Biochem. J.,* 219, 911, 1984.

166. **Butterwith, S. C., Martin, A., and Brindley, D. N.,** Can phosphorylation of phosphatidate phosphohydrolase by a cyclic AMP-dependent mechanism regulate its activity and subcellular distribution and control hepatic glycerolipid synthesis? *Biochem. J.,* 222, 487, 1984.

167. **Butterwith, S. C., Martin, A., Cascales, C., Mangiapane, E. H., and Brindley, D. N.,** Regulation of triacylglycerol synthesis by the translocation of phosphatidate phosphohydrolase from the cytosol to the membrane-associated compartment, *Biochem. Soc. Trans.,* 13, 158, 1985.

168. **Martin, A., Hopewell, R., Martin-Sanz, P., Morgan, J. E., and Brindley, D. N.,** Relationship between the displacement of phosphatidate phosphohydrolase from the membrane-associated compartment by chlorpromazine and the inhibition of the synthesis of triacylglycerol and phosphatidylcholine in rat hepatocytes, *Biochim. Biophys. Acta,* 876, 581, 1986.

169. **Martin-Sanz, P., Hopewell, R., and Brindley, D. N.**, Long-chain fatty acids and their acyl-CoA esters cause the translocation of phosphatidate phosphohydrolase from the cytosolic to the microsomal fraction of rat liver, *FEBS Lett.*, 175, 284, 1984.

170. **Martin-Sanz, P., Hopewell, R., and Brindley, D. N.**, Spermine promotes the translocation of phosphatidate phosphohydrolase from the cytosol to the microsomal fraction of rat liver and it enhances the effects of oleate in this respect, *FEBS Lett.*, 179, 262, 1985.

171. **Hopewell, R., Martin-Sanz, P., Martin, A., Saxton, J., and Brindley, D. N.**, Regulation of the translocation of phosphatidate phosphohydrolase between the cytosol and the endoplasmic reticulum of rat liver, *Biochem. J.*, 232, 485, 1985.

172. **Moller, F. and Hough, M. R.**, Effects of salts of membrane binding and activity of adipocyte phosphatidate phosphohydrolase, *Biochim. Biophys. Acta*, 711, 521, 1982.

173. **Jamdar, S. C. and Osborne, L. J.**, Glycerolipid biosynthesis in rat adipocyte tissue. II. Effects of polyamines of Mg^{2+}-dependent phosphatidate phosphohydrolase, *Biochim. Biophys. Acta*, 752, 79, 1983.

174. **Jamdar, S. C. and Osborne, L. J.**, Glycerolipid biosynthesis in rat adipose tissue. IX. Activation of diglyceride acyltransferase by spermine, *Enzyme*, 28, 387, 1983.

175. **Hong, K., Schuber, F., and Papahadjopoulos, D.**, Polyamines. Biological modulators of membrane fusion, *Biochim. Biophys. Acta*, 732, 469, 1983.

176. **Schuber, F., Hong, K., Duzkunes, N., and Papahadjopoulos, D.**, Polyamines as modulators of membrane fusion: aggregation and fusion of liposomes, *Biochemistry*, 22, 6134, 1983.

177. **Chung, L., Kaloyanides, G., McDaniel, A., McLaughlin, A., and McLaughlin, S.**, Interaction of gentamycin and spermine with bilayer membranes containing negatively charged phospholipids, *Biochemistry*, 24, 442, 1985.

178. **Tadolini, B., Cabrini, L., Varani, E., and Sechi, A. M.**, Spermine binding and aggregation of vesicles of different phospholipid composition, *Biogenic Amines*, 3, 87, 1985.

179. **Theoharides, T. C.**, Polyamines, spermidine and spermine as modulators of calcium-dependent immune processes, *Life Sci.*, 27, 703, 1980.

180. **Canellakis, Z. N. and Theoharides, T. C.**, Stimulation of ornithine decarboxylase synthesis and its control by polyamines in regenerating rat liver and cultured rat hepatoma cells, *J. Biol. Chem.*, 251, 4436, 1976.

181. **Hölttä, E.**, Immunochemical demonstration of increased accumulation of ornithine decarboxylase in rat liver after partial hepatectomy and growth hormone induction, *Biochim. Biophys. Acta*, 399, 420, 1975.

182. **Tabor, C. W. and Tabor, H.**, 1,4-Diaminobutane (putrescene), spermidine and spermine, *Annu. Rev. Biochem.*, 45, 285, 1976.

183. **Williams-Ashman, H. G. and Canellakis, Z. N.**, Polyamines in mammalian biology and medicine, *Perspect. Biol. Med.*, 22, 421, 1979.

184. **Sturton, R. G. and Brindley, D. N.**, Factors controlling the activities of phosphatidate phosphohydrolase and phosphatidate cytidylyltransferase, *Biochem. J.*, 162, 25, 1977.

185. **Means, A. R.**, Calmodulin: properties, intracellular localization and multiple roles in cell regulation, *Recent Prog. Hormone Res.*, 37, 333, 1981.

186. **Rainteau, D., Wolfe, C., Bereziat, G., and Polonovski, J.**, Binding of a spin labelled chlorpromazine analogue to calmodulin, *Biochem. J.*, 221, 659, 1984.

187. **Roufogalis, B. D.**, Phenothiazine antagonism of calmodulin: a structural-nonspecific interaction, *Biochem. Biophys. Res. Commun.*, 98, 607, 1981.

188. **Mori, T., Takai, Y., Minakuchi, R., Yus, B., and Nishizuka, Y.**, Inhibitory action of chlorpromazine, dibucaine and other phospholipid-interacting drugs on calcium-activated phospholipid dependent protein kinase, *J. Biol. Chem.*, 255, 8378, 1980.

189. **Kuo, J. F., Andersson, R. G. G., Wise, B. C., MacKerlova, L., Salomonsson, L., Brackett, N. L., Kotoh, N., Shoji, M., and Wren, R. W.**, Calcium dependent protein kinase: widespread occurrence in various tissues and phyla of the animal kingdom and comparison of effects of phospholipid, calmodulin and trifluoperazine, *Proc. Natl. Acad. Sci. U.S.A.*, 77, 7039, 1980.

190. **Schatzman, R. C., Wise, B. C., and Kuo, J. F.**, Phospholipid-sensitive calcium-dependent protein kinase: inhibition by antipsychotic drugs, *Biochem. Biophys. Res. Commun.*, 98, 669, 1981.

191. **Kraft, A. S. and Anderson, W. B.**, Phorbol esters increase the amount of Ca^{2+}-phospholipid dependent protein kinase associated with plasma membrane, *Nature (London)*, 301, 621, 1983.

192. **Wooten, M. W. and Wrenn, R. W.**, Phorbol ester induces intracellular translocation of phospholipid/Ca^{2+}-dependent protein kinase and stimulates amylase secretion in isolated pancreatic acini, *FEBS Lett.*, 171, 183, 1984.

193. **Berglund, L., Björkhem, I., and Einarsson, K.**, Apparent phosphorylation-dephosphorylation of soluble phosphatidic acid phosphatase in rat liver, *Biochem. Biophys. Res. Commun.*, 105, 288, 1982.

194. **Björkhem, I., Angeln, B., Backman, L., Liljequist, L., Nikell, K., and Einarsson, K.**, Triglyceride metabolism in human liver: studies on hepatic phosphatidic acid phosphatase in obese and nonobese subjects, *Eur. J. Clin. Invest.*, 14, 233, 1984.

195. **Berglund, L., Angelin, B., Björkhem, I., and Einarsson, K.,** Regulation of triglyceride biosynthesis: studies on hepatic phosphatidic acid phosphatase, in *Treatment of Hyperlipoproteinemia,* Carlson, L. A. and Olsson, A. G., Eds., Raven Press, New York 1984, 99.

196. **Soler-Argilaga, C., Russel, R. L., and Heimberg, M.,** Enzymatic aspects of the reduction of microsomal glycerolipid biosynthesis after perfusion of the liver with dibutyryladenosine-3′,5′ monophosphate, *Arch. Biochem. Biophys.,* 190, 367, 1978.

197. **Pelech, S. L., Pritchard, P. H., Brindley, D. N., and Vance, D. E.,** Fatty acids reverse the cyclic AMP inhibition of triacylglycerol and phosphatidylcholine synthesis in rat hepatocytes, *Biochem. J.,* 216, 129, 1983.

198. **Houslay, M. D. and Marchmont, R. J.,** The insulin-stimulated cyclic AMP phosphodiesterase binds to a single class of protein sites on the liver plasma membrane, *Biochem. J.,* 198, 703, 1981.

199. **Brownsey, R. W. and Denton, R. M.,** Evidence that insulin activates fat cell acetyl-CoA carboxylase by increased phosphorylation at a specific site, *Biochem. J.,* 202, 77, 1982.

200. **Hübscher, G., Brindley, D. N., Smith, M. E., and Sedgwick, B.,** Stimulation of biosynthesis of glyceride, *Nature (London),* 216, 449, 1967.

201. **Johnston, J. M., Rao, G. A., Lowe, P. A., and Schwarz, B. E.,** The nature of the stimulatory role of the supernatant fraction of triglyceride synthesis by the α-glycerophosphate pathway, *Lipids,* 2, 14, 1967.

202. **Brindley, D. N., Smith, M. E., Sedgwick, B., and Hübscher, G.,** The effect of unsaturated fatty acids and the particle-free supernatant on the incorporation of palmitate into glycerides, *Biochim. Biophys. Acta,* 144, 285, 1967.

203. **Roncari, D. A. K. and Mack, E. Y. W.,** Stimulation of triacylglycerol synthesis in mammalian liver and adipose tissue by two cytosolic compounds, *Biochem. Biophys. Res. Commun.,* 67, 790, 1975.

204. **Roncari, D. A. K. and Mack, E. Y. W.,** Purification of cytosolic proteins that stimulate triacylglycerol synthesis, *Can. J. Biochem.,* 59, 944, 1981.

205. **Pelech, S. L. and Vance, D. E.,** Regulation of phosphatidylcholine biosynthesis, *Biochim. Biophys. Acta,* 779, 217, 1984.

206. **Vance, D. E. and Pelech, S. L.,** Enzyme translocation in the regulation of phosphatidylcholine biosynthesis, *Trends Biochem. Sci.,* 9, 17, 1984.

207. **Pelech, S. L., Pritchard, P. H., Brindley, D. N., and Vance, D. E.,** Fatty acids promote translocation of CTP:phosphocholine cytidylyltransferase to the endoplasmic reticulum and stimulate hepatic phosphatidylcholine synthesis, *J. Biol. Chem.,* 258, 6782, 1983.

208. **Weinhold, P. A., Rounsifer, M. E., Williams, S. E., Brubaker, P. G., and Feldman, D. A.,** CTP:phosphorylcholine cytidylyltransferase in rat lung. The effect of free fatty acids on the translocation of activity between microsomes and cytosol, *J. Biol. Chem.,* 259, 10315, 1984.

209. **Feldman, D. A., Rounsifer, M., and Weinhold, P. A.,** The stimulation and binding of CTP:phosphorylcholine cytidylyltranferase by phosphatidylcholine-oleic acid vesicles, *Biochim. Biophys. Acta,* 833, 429, 1985.

210. **Wilson, J. E.,** Brain hexokinase, the prototype ambiquitous enzyme, *Curr. Topics Cell. Regul.,* 16, 1, 1980.

211. **Lim, P., Cornell, R., and Vance, D. E.,** The supply of both CDP-choline and diacylglycerol can regulate the rate of phosphatidylcholine synthesis in HeLa cells, *Biochem. Cell. B,* 64, 692, 1986.

212. **Pelech, S. L., Jetha, F., and Vance, D. E.,** Trifluoperazine and other anaesthetics inhibit rat liver CTP:phosphocholine cytidylyltransferase, *FEBS Lett.,* 158, 89, 1983.

213. **Pelech, S. L. and Vance, D. E.,** Trifluoperazine and chlorpromazine inhibit phosphatidylcholine biosynthesis and CTP:phosphocholine cytidylyltransferase in HeLa cells, *Biochim. Biophys. Acta,* 795, 441, 1984.

214. **Pelech, S. L., Paddon, H. B., and Vance, D. E.,** Phorbol esters stimulate phosphatidylcholine biosynthesis by translocation of CTP: phosphocholine cytidylyltransferase from cytosol to microsomes, *Biochim. Biophys. Acta,* 795, 447, 1984.

215. **Cornell, R. and Vance, D. E.,** Translocation of CTP: phosphocholine cytidylyltransferase from cytosol to membranes in Hela cells: stimulation by fatty acid, fatty alcohol, mono- and diacylglycerol, *Biochim. Biophys. Acta,* 919, 26, 1987.

216. **Caras, I. and Shapiro, B.,** Partial purification and properties of microsomal phosphatidate phosphohydrolase from rat liver, *Biochim. Biophys. Acta,* 409, 201, 1975.

217. **Hall, M., Taylor, S. J., and Saggerson, E. D.,** Persistent activity modification of phosphatidate phosphohydrolase and fatty acyl-CoA synthetase on incubation of adipocytes with the tumour promoter 12-O-tetradecanoylphorbol 13-acetate, *FEBS Lett.,* 179, 351, 1985.

218. **van Heusden, G. P. and van den Bosch, H.,** The influence of exogenous and of membrane-bound phosphatidate concentration on the activity of CTP:phosphatidate cytidylyltransferase and phosphatidate phosphohydrolase, *Eur. J. Biochem.,* 84, 405, 1978.

219. **Brindley, D. N., Bowley, M., Sturton, R. G., Pritchard, P. H., Cooling, J., and Burditt, S. L.,** The effects of amphiphilic compounds on phosphatidate metabolism, *Adv. Exp. Med. Biol.,* 101, 227, 1978.

220. **Pikkukangas, A. H., Väänänen, R. A., Savolainen, M. J., and Hassinen, I. E.,** Precursor supply and hepatic enzyme activities as regulators of triacylglycerol synthesis in isolated hepatocytes and perfused liver, *Arch. Biochem. Biophys.*, 217, 216, 1982.

221. **Haagsman, H. P., de Haas, C. G. M., Geelen, M. J. H., and van Golde, L. M. G.,** Regulation of triacylglycerol synthesis in the liver. A decrease in diacylglycerol acyltransferase activity after treatment of isolated rat hepatocytes with glucagon, *Biochim. Biophys. Acta*, 664, 74, 1981.

222. **Haagsman, H. P., de Haas, C. G. M., Geelen, M. J. H., and van Golde, L. M. G.,** Regulation of triacylglycerol synthesis in the liver, *J. Biol. Chem.*, 257, 10593, 1982.

223. **Haagsman, H. P. and van Golde, L. M. G.,** Synthesis and secretion of very low density lipoproteins by isolated rat hepatocytes in suspension: role of diacylglycerol acyltransferase, *Arch. Biochem. Biophys.*, 208, 395, 1981.

224. **Nikkilä, E. A. and Kekki, M.,** Plasma triglyceride transport kinetics in diabetes mellitus, *Metabolism*, 22, 1, 1973.

225. **Brindley, D. N.,** Metabolism of triacylglycerols, in *Biochemistry of Lipids and Membranes*, Vance, D. E. and Vance, J., Eds., Addison-Wesley, Reading, Mass., 1985, chap. 7.

226. **Cryer, A.,** Tissue lipoprotein lipase activity and its action in lipoprotein metabolism, *Int. J. Biochem.*, 13, 525, 1981.

227. **Sweeney, D. and Brindley, D. N.,** unpublished work.

228. **Lawson, N., Mangiapane, E. H., and Brindley, D. N.,** unpublished work.

229. **Martin, A., Hales, P., and Brindley, D. N.,** unpublished work.

230. **Cornell, R. and Vance, D. E.,** Binding of CTP:phosphocholine cytidylyltransferase to large unilamellar vesicles, *Biochim. Biophys. Acta*, 919, 37, 1987.

Chapter 3

PHOSPHATIDATE PHOSPHOHYDROLASE ACTIVITY IN ADIPOSE TISSUE

E. D. Saggerson

TABLE OF CONTENTS

I. INTRODUCTION

Adipose tissue has evolved, particularly in mammals, for storage of triacylglycerol in specialized cells, the adipocytes. Two types of adipose tissue are recognized according to their color and the form of their adipocytes: (1) white, or unilocular adipose tissue and (2) brown, or multilocular adipose tissue. There is no single definable white adipose tissue organ in mammals, rather, the tissue is located diffusely in a number of distinct anatomical sites (e.g., subcutaneous, mesenteric, interscapular, retroperitoneal). Most biochemical studies have been performed on the large and accessible depots of intra-abdominal white adipose tissue located in the male rat or mouse adjacent to the epididymis and testis and in the female alongside the uterine horns. It is now recognized that white adipose tissue is metabolically very active and well vascularized. There is adrenergic innervation to both vasculature and to adipocytes themselves with, additionally, cellular junctions permitting communication between cells. White adipocytes are generally large cells varying in size from 10 to 120 μm in diameter and their most conspicuous feature is the single large lipid droplet which in the fed state fills most of the cell. Surrounding this lipid droplet is a thin peripheral rim containing both the nucleus and the cytoplasm which has an abundance of smooth endoplasmic reticulum. The reader is referred to other specialist reviews for further information on the morphology of white adipose tissue.[1-3] White adipose tissue can act as a mechanical buffer to protect delicate internal organs and to provide thermal insulation but it is its role as an energy store that particularly interests the biochemist and the physician. In essence, its role is the storage of long chain fatty acids as triacylglycerols in times of energy surfeit and the mobilization of fatty acids out of this triacylglycerol store in times of anticipated or actual energy demand.

A certain amount of general ''turnover'' of these stores probably occurs continuously but whether the tissue is bringing about net deposition or net mobilization will depend upon the relative activities of the lipolytic and esterification processes, the phosphohydrolase playing a prominent role in the latter. As will be described, these two key opposing adipose tissue processes are generally subject to inverse control by endocrine, neuroendocrine, and pharmacological factors, but they may also interact upon each other to some extent.

Brown adipose tissue is mainly located in the interscapular and axillary regions. Its color is due to its rich vascular supply and the presence of numerous large mitochondria within the adipocytes. The cells range in diameter from 8 to 60 μm and contain numerous lipid droplets in close association with the endoplasmic reticulum and the mitochondria. The tissue is specialized for production of heat, mainly derived from a high rate of uncoupled respiration of mobilized fatty acids. Nonshivering thermogenesis in brown adipose tissue is particularly important in mammals during early postnatal life, during cold exposure, and on arousal from hibernation. The tissue may also be involved in diet-induced thermogenesis in some animals. Irrespective of the cause, the main stimulus for thermogenesis appears to be noradrenaline released from the dense sympathetic nerve supply to the tissue. The reader is referred to specialist texts for further information on brown adipose tissue structure and function[3-7] but it can be stated quite simply that although the processes whereby thermogenesis is activated are relatively well understood we have little knowledge about triacylglycerol synthesis, its control, and the role of the phosphohydrolase in brown adipose tissue.

II. THE CONTROL OF ADIPOSE TISSUE TRIACYLGLYCEROL SYNTHESIS AND MOBILIZATION

The interrelated regulation of these two processes is discussed in this section in order to set the physiological scene before describing the properties and regulation of the phosphohydrolase.

A. Triacylglycerol Synthesis
1. The Pathway in White Adipose Tissue
Figure 1 shows a generalized scheme of the process giving some indication of the compartmentation of its components. Glycerolipid synthesis in adipose tissue is essentially dedicated to the formation of triacylglycerols with only a minute fraction of the pathway flux giving rise to phospholipid products. Unlike liver[8,9] the microsomal form of glycerolphosphate acyltransferase predominates over the mitochondrial activity by a factor of at least five under most assay conditions.[10-12] In addition, dihydroxyacetone phosphate acyltransferase activity is small compared to that of glycerolphosphate acyltransferase[11,13,14] and is essentially confined to the microsomal fraction[11,15] with very little N-ethylmaleimide-insensitive dihydroxyacetone phosphate acyltransferase[11] which in liver may be found in either mitochondria or peroxisomes.[9,16,17] Monoacylglycerol phosphate acyltransferase[11] and diacylglycerol acyltransferase[11,18] activities are also essentially confined to microsomes. The phosphatidate phosphodydrolase is found in both soluble and microsomal fractions,[11,19,21] and, as discussed in Section VI appears to be ambiquitous. The specific activities of these four microsomal acyltransferases in microsomes from white adipocytes are in general at least five to ten times higher than those observed in liver[9,22,26] and reflect the very high capacity in adipocytes of the endoplasmic reticulum triacylglycerol synthetic pathway. Under optimum conditions of supply of glycerolphosphate and acyl substrates (i.e., cells incubated

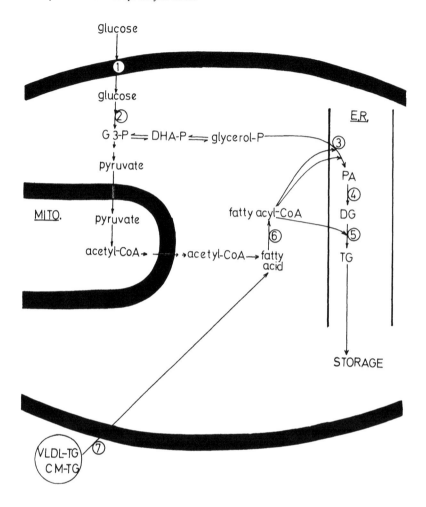

FIGURE 1.   General scheme outlining pathways for the synthesis of fatty acids and triacylglycerol in white adipose tissue. Steps which are particularly discussed in the text are
(1) glucose transport; (2) phosphofructokinase, (3) glycerolphosphate acyltransferase, (4)
phosphatidate phosphohydrolase, (5) diacylglycerol acyltransferase, (6) fatty acyl-CoA synthetase, and (7) lipoprotein lipase. Abbreviations used are G3-P, glyceraldehyde 3-phosphate;
DHA-P, dihydroxyacetone phosphate; PA, phosphatidate; DG, diacylglycerol; TG, triacylglycerol; mito, mitochondria; CM, chylomicron; VLDL, very low density lipoprotein; ER,
endoplasmic reticulum.

with saturating level of glucose, insulin, and palmitate) rat adipocytes have been observed[27,28]
to esterify 30 μmol of palmitate per hour per 100 μg of adipocyte DNA which represents
an approximate 5% increase in cell triacylglycerol content per hour.

White adipose tissues mainly acquire long chain fatty acids for esterification and storage
via the action of lipoprotein lipase on plasma lipoprotein triacylglycerols. These may be in
chylomicrons representing the products of digestion and absorption of dietary fat or in very
low density lipoproteins allowing fatty acids synthesized *de novo* in the liver to be distributed
to the storage depots. In addition, a certain proportion of stored fatty acid may be synthesized
de novo within the adipocyte, mainly from blood glucose and lactate.[29-32] The contribution
of this later process to the aquisition of fatty acids by the adipocyte is variable being both
species- and age-dependent.

White adipose tissue contains little glycerokinase activity and therefore free glycerol is
not important as a precursor of the triacylglycerol glycerol moiety. Blood glucose however

Table 1
THE EFFECT OF EXOGENOUS FATTY ACID ON TOTAL GLUCOSE UTILIZATION AND ON GLYCEROLIPID SYNTHESIS BY RAT ADIPOCYTES

Glucose conc. (mM)	Insulin	Palmitate/albumin molar ratio	Total glucose utilization (μmol.hr^{-1}.100 μg cell DNA^{-1})	Glucose utilization for synthesis of glyceride glycerol (%)
5	+	0	9.3	6.9
5	+	0.67	14.0	13.7
5	+	1.34	15.7	19.3
5	+	2.7	15.9	27.4
5	+	5.4	16.3	30.2
1	−	0	2.2	11.7
1	−	1.34	3.7	42.6
1	−	5.4	4.9	45.2

Note: The data are derived from experiments of Saggerson[28] in which rat adipocytes were incubated for 1 hr in media containing 5% (w/v) albumin with or without insulin (20 munit/mℓ) and the indicated concentration of [U-^{14}C]-glucose.

Table 2
COMPARISON OF RATES OF GLYCEROLIPID SYNTHESIS BY RAT ADIPOCYTES — DEPENDENCE ON GLYCEROL PHOSPHATE PRECURSOR AND EFFECT OF NUTRITIONAL STATE[27,281]

Nutritional state	[U-^{14}C]-glucose, pyruvate, or lactate	Additions to incubations		Rate of glycerolipid synthesis as μgatoms of ^{14}C incorporated into glyceride glycerol.hr^{-1}. 100 μg cell DNA^{-1}
		Insulin (20 munit/mℓ)	Palmitate/albumin molar ratio	
Fed	Pyruvate (5 mM)	+	5.4	1.6
	Lactate (5 mM)	+	5.4	0.3
	Glucose (1 mM)	+	5.4	23.0
	Glucose (5 mM)	+	5.4	29.8
	Glucose (1 mM)	−	5.4	13.0
	Glucose (5 mM)	−	5.4	24.7
	Glucose (5 mM)	−	0	1.8
	Glucose (5 mM)	+	0	3.7
	Glucose (5 mM)	+	4.9	23.7
24-hr starved	Glucose (5 mM)	+	4.9	16.3
72-hr starved	Glucose (5 mM)	+	4.9	7.8

is most certainly a very important precursor of glycerolphosphate and hence glyceride glycerol. Under certain circumstances (Table 1) formation of glyceride glycerol becomes a very significant proportion of the total utilization of glucose by the adipocyte. Although white adipocytes contain the appropriate enzyme activities,[27,33-36] 3-carbon precursors such as lactate and pyruvate are very poor precursors of glyceride glycerol and can alone support only a small amount of fatty acid esterification (Table 2).[27,29,31,37,38] As also shown in Table 2 storage of fatty acids *synthesized de novo* only utilizes approximately 10 to 15% of the capacity of the esterification pathway in incubated adipocytes even when rates of fatty acid synthesis are maximal (5 mM glucose and insulin present). The rate of esterification is greatly enhanced in vitro either by stimulating lipolysis to provide a source of nonesterified fatty acids (see Section II.B) or by addition of an exogenous fatty acid such as palmitate (Table 2).[27-29,37,39,41] In vivo, the latter situation will be replaced by the action of lipoprotein lipase to provide the source of fatty acyl substrates.

2. Regulation of Triacylglycerol Synthesis in White Adipose Tissue

This process in white fat is controlled both by the availability of the pathway precursors and by controls within the triacylglycerol synthesis pathway itself. In liver it has been proposed that control of glycerolipid synthesis is significantly influenced in a manner secondary to events regulating the alternative fate of fatty acids in that tissue, namely β-oxidation.[42-44] It is most unlikely that this is the case in white adipose tissue since rates of fatty acid oxidation are extremely low relative to esterification[40] and therefore changes in oxidation must have negligible effect on esterification.

a. Control of Precursor Availability
i. Fatty Acids

In vivo, the activity of lipoprotein lipase is a major determinant of the rate at which adipose tissue extracts fatty acids from circulating lipoprotein triacylglycerols. In times of caloric sufficiency this activity is elevated whereas caloric shortage or stress states will be associated with its decrease. Lipoprotein lipase in the adipose tissue capillary bed is synthesized within and secreted by the adipocytes[45-47] and this process is decreased in starvation but restored on refeeding,[48,49] it is decreased in diabetes but restored by insulin,[50-52] and it is decreased during lactation.[53,54] Insulin appears to be particularly important in the maintenance of active lipoprotein lipase enzyme in the tissue[55-59] and this effect of insulin is opposed by various lipolytic agents including the catecholamines, glucagon, corticotropin, methylxanthines, and dibutyryl cAMP.[60-64] The properties of lipoprotein lipase and its control are reviewed more fully elsewhere.[47,65] The associated, but less significant contribution to fatty acid supply, namely synthesis *de novo*, within the adipocyte is regulated in a parallel manner to the activity of lipoprotein lipase being in the short term, substantially increased by insulin[39,41,66,67] and generally, decreased by catecholamines.[39,41,67] In the long term the process is considerably decreased by fasting and in diabetes.[68] These changes appear to mainly center around the control of glucose transport (see Section II.A.2.a.ii), pyruvate dehydrogenase, and acetyl-CoA carboxylase.[29,69-71]

It would be erroneous to assume that supply of fatty acyl precursors to the triacylglycerol synthesis pathway must automatically be diminished in times of caloric insufficiency or stress when lipoprotein lipase and synthesis of fatty acids *de novo* are shut down. Increased levels of lipolytic hormones and/or diminished levels of insulin cause triacylglycerol mobilization (see Section II.B) and in vitro a certain proportion of mobilized fatty acid is recycled back into triacylglycerol when adipocytes are incubated in closed systems.[28,39,41] In vivo blood flow will obviously diminish this but there is nevertheless evidence for triacylglycerol/fatty acid substrate cycling.[72] It is the author's contention however that certain regulatory mechanisms act to minimize this potentially wasteful cycling (see Section II.A. 2.b). Among such mechanisms may be catecholamine-mediated stimulation of fatty acid transport out of the adipocyte.[73,74]

ii. Glycerolphosphate

The intracellular supply of this crucial precursor is extremely dependent upon metabolism of glucose to the triose phosphates (see Section II.A.1). In this process there are two steps clearly defined as being regulatory, namely the glucose transporter and phosphofructokinase.

Transport of glucose into adipocytes is a facilitated diffusion process that is considerably increased by insulin.[75] This is reflected in enhancement by the hormone of transport of

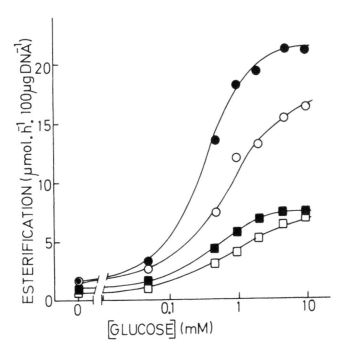

FIGURE 2. Dependence of triacylglycerol synthesis by rat adipocytes on glucose. The data are derived from Harper and Saggerson[40] and show rates of esterification when cells are incubated with 0.75 mM [1-^{14}C]-palmitate, albumin (21 mg/mℓ) with (closed symbols) or without (open symbols) insulin (20 munit/mℓ). (○,●) Cells from fed animals; (□,■) cells from 48 hr-starved animals.

nonmetabolizable analogs[76-85] and increased rates of glucose utilization[39,66,67,86-88] by adipose tissue preparations. This response appears to be achieved by insulin-stimulated recruitment to the plasma membrane of glucose transporters from an intracellular store[69,82,84,89-91] and probably also by modification of the properties of intrinsic transporter molecules[84,85,92-94] in the plasma membrane. Other physiologically relevant stimulators of glucose transport are long chain fatty acids[95] and adenosine,[84,96,97] a paracrine agent that affects other aspects of adipose tissue metabolism (see Section II.B). Adipose tissue phosphofructokinase is implicated as a regulatory site on the basis of the tissue contents of its reactants[39,88] and because of its kinetic properties which include inhibition by glycerol-phosphate, ATP and citrate, sigmoidal kinetics for its substrate fructose 6-phosphate, and stimulation by cAMP, AMP, long chain fatty acids, and fructose 2,6-bisphosphate.[98,102] Increased transport of glucose in response to insulin and fatty acid supply is accompanied by an increased rate of glycolysis,[39,66,67,88,101] an increase in the fructose 6-phosphate substrate for phosphofructokinase,[39,88,101,103] a persistent increase in the V_{max} of phosphofructokinase,[99,101] and either an increase[103] or small decrease[101] in the tissue content of fructose 2,6-bisphosphate. The end result is an increase in the level of glycerolphosphate,[39,86-88,101] and, when fatty acid is provided as cosubstrate, insulin and glucose have interactive effects to increase triacylglycerol synthesis[40,104] (see Table 1 and Figure 2).

When adipose tissue is stimulated by lipolytic agents re-esterification of fatty acids also means that there is increased demand for glycerolphosphate. Generally[83,84,96,105-107] but not exclusively,[83,84,108,109] there is a decrease in the V_{max} of the glucose transporter under these conditions. However, net utilization of glucose is invariably increased[28,39,67,88,110-114] and this appears most likely to be achieved by activation of phosphofructokinase. In the presence of lipolytic agents, adipose tissue phosphofructokinase sustains an increase in glycolytic flux in the face of a decrease in the level of its substrate fructose 6-phosphate.[39,103] Possible mechanisms that are involved are activation by fatty acids,[99,100] phosphorylation of the enzyme by cAMP-dependent protein kinase[115] resulting in decreased sensitivity to ATP inhibition,[102] and increased sensitivity to fructose 2,6-bisphosphate activation.[100,120] At the same time there is a fall in ATP level, an increase in the level of the activator AMP, and a decrease in glycerol phosphate concentration.[39,86-88] The level of fructose 2,6-bisphosphate is either unchanged[100] or decreased.[103] The overall outcome of these complex interactions is that some re-esterification is permitted under lipolytic conditions but this is probably limited by the fall in glycerol phosphate level and indirectly by the control of the glucose transporter.

Provision of glycerol phosphate is also likely to be decreased in longer-term situations, particularly through changes in the activity or availability of the glucose transporter. These states include dexamethasone treatment,[116,117] fasting,[107,118] diabetes,[119] age and obesity,[120] and high fat, low carbohydrate diets.[121] In some instances these changes are known to correlate with decreased triacylglycerol synthesis (see Section II.B).

b. Control within the Triacylglycerol Synthesis Pathway

Our knowledge of this is still poor. In the short term, under appropriate experimental conditions, lipolytic agents such as adrenaline or dibutyryl cAMP can be shown to decrease esterification in incubated fat pieces or adipocytes.[29,122,123] Exposure of adipocytes to lipolytic agents results in rapid and relatively persistent decreases in the activities of glycerolphosphate acyltransferase,[12,14,124,123] dihydroxyacetonephosphate acyltransferase,[14] fatty acyl-CoA synthetase,[126,127] phosphatidate phosphohydrolase,[128,131] and diacylglycerol acyltransferase[131,132] (also see Section IV.A.1). Insulin has been shown to rapidly increase the activity of fatty acyl-CoA synthetase[133] although this finding has not been confirmed by others.[126,127-134] On the other hand, insulin blocks catecholamine-induced decreases in these activities[125,126,129,132] and rapidly reverses the catecholamine effects on glycerolphosphate acyltransferase,[12] phosphatidate phosphohydrolase,[129] and fatty acyl-CoA synthetase.[127] The mechanisms underlying such effects are unknown. It was thought[135] that phosphorylation by cAMP-dependent protein kinase might mediate the inactivation of glycerolphosphate acyltransferase by catecholamines but this now appears unlikely.[12,136] It is suggested by the author that rapid down-regulation of these enzyme activities under lipolytic conditions is important to restrain wasteful recycling of mobilized fatty acids back into the triacylglycerol stores. The rat adipocyte esterification system is also subject to long-term regulation. Starvation results in decreased incorporation of [^{14}C]-glucose into glyceride glycerol[37] or of [^{14}C]-palmitate into triacylglycerol[40] (see Table 1 and Figure 2). In accord with this, rat adipose tissue glycerolphosphate acyltransferase activity is decreased in starvation,[21,137,139] in streptozotocin-diabetes,[21] and in hypothyroidism.[21]

Parallel changes in the phosphohydrolase activity are discussed in Volume II, Chapter 6, Section V.

3. Brown Adipose Tissue

Our very sparse knowledge of the esterification process and its regulation in this tissue can be summarized as follows. [^{14}C]-glucose is able to act as a precursor for the glycerol moiety of rat and rabbit brown fat glycerides[140-144] and insulin increases the rate of this process, presumably through stimulation of glucose uptake.[145,146] In addition, the process in neonatal rabbit tissue is increased by noradrenaline and in cold adaptation.[143,144] Unlike white adipose tissue brown fat contains high glycerokinase activity[143,147] and therefore glycerol must also be considered as an immediate precursor for glyceride synthesis. A further difference from white adipose tissue is that the esterification pathway in brown fat has to

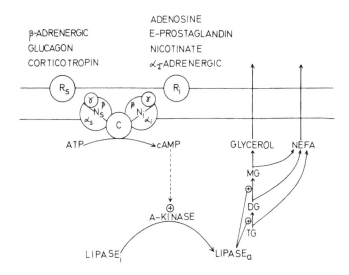

FIGURE 3. General scheme for the dual regulation of adipose tissue lipolysis. This shows that phosphorylation and activation of the hormone-sensitive lipase by cAMP-dependent protein kinase is under dual control by agonists which either stimulate or inhibit adenylate cyclase. These each have plasma membrane receptors (R_s or R_i for stimulators or inhibitors respectively) which couple to the catalytic unit of the cyclase via the guanine nucleotide-binding proteins N_s and N_i, α, β, V indicate the subunits of the N_s and N_i proteins. C represents the catalytic entity of adenylate cyclase. Inhibition via the α_2-adrenoceptor does not appear to occur in the rat but is observed in man and the hamster. Not shown in the scheme is the effect of insulin which also inhibits the process via an unknown mechanism. Abbreviations used: TG, DG, and MG = tri-, di-, and monoacylglycerol respectively; NEFA, nonesterified fatty acids.

compete with the very active β-oxidation process for the available fatty acid. In hamster, brown adipocytes esterification of [^{3}H]-oleate takes place at a rate that is approximately 20% of the rate of β-oxidation on a mole/mole basis and esterification still proceeds even when β-oxidation is stimulated by noradrenaline.[148] It is not known how the balance between these two competing processes is regulated in this tissue. It is reported that fatty acid esterification in hamster brown fat occurs via both the monoacylglycerol and the glycerol-phosphate pathways[149] whereas it has been reported that only the glycerol-phosphate pathway is operative in rabbit cells.[150] Besides phosphatidate phosphohydrolase (see Section III), the only enzyme of the esterification pathway to be investigated in brown fat is fatty acyl-CoA synthetase which is present at high specific activity[151,152] localized mainly in the mitochondrial outer membrane,[151] little activity being found in the endoplasmic reticulum which is reported as being sparse in brown fat.[152]

B. Triacylglycerol Mobilization

The rate of this process in white adipose tissue is dependent on the interplay between a number of factors both in the short and long term. Much of the work in this area has been reviewed elsewhere[154-160] from various standpoints. The central feature of lipolysis is the hormone-sensitive lipase which is rate-limiting for the process and catalyzes hydrolysis of triacylglycerol to diacylglycerol and diacylglycerol to monoacylglycerol (Figure 3). The hydrolysis of the resulting monoacylglycerol is mainly catalyzed by a separate monoacylglycerol lipase. Hormone-sensitive lipase is activated by cAMP-dependent protein kinase[161-171] and it is now established that this is due to the phosphorylation of the enzyme[172-175] at a single serine residue.[176,177] Dephosphorylation and deactivation of the lipase can be achieved by cellular protein phosphatases of types 1A, 2A, and 2C.[160]

A variety of lipolytic and antilipolytic agents acting through plasma membrane receptors[178] are able to influence the activity state of adenylate cyclase and thereby the cellular level of cAMP and the activity of the cAMP-dependent protein kinase. The most frequently considered lipolytic hormones are adrenaline and noradrenaline (acting at β-adrenoceptors), corticotropin, and glucagon all of whose receptors are coupled to the catalytic unit of the adenylate cyclase by the guanine nucleotide, binding protein N_s or G_s.[179] On the other hand, receptors for antilipolytic agents such as adenosine, E-series prostaglandins, and nicotinate are coupled to the cyclase by the distinct guanine nucleotide-binding protein entitled N_i or G_i.[179] The qualitative relationship between cAMP-dependent protein kinase activity and lipolysis has been described for incubated white adipocytes and is found to be independent of combinations of the above lipolytic or antilipolytic agents tested.[180] This finding suggests that all these agents act on the lipolytic process exclusively by altering adenylate cyclase activity and thus cellular cAMP concentration.[180] The responsiveness or sensitivity of this dual regulation system to both the stimulatory and the inhibitory agonists changes in the longer term with alterations in physiological state. Thus, hypothyroidism,[181-189] adrenalectomy,[190-194] and lactation[195-197] are associated with decreased responsiveness to lipolytic hormones and increased responsiveness to the antilipolytic agents. An opposite situation is encountered in fasting[186,189,198-201] and diabetes[202-204] where responsiveness to stimulatory and inhibitory agonists is increased and decreased respectively. It is probable that these longer term changes primarily reflect postreceptor alterations in the N_s and N_i-mediated coupling processes. Insulin is also a physiologically important antilipolytic agent that decreases the phosphorylation state of the hormone-sensitive lipase.[160,205] However, this receptor-mediated action of insulin does not appear to be mediated through N_i.[206] While part of this action of insulin may be attributable to lowering of cellular cAMP level[207-210] and cAMP-dependent protein kinase activity, part of the hormone's antilipolytic effect is cAMP-independent[211-214] and is tentatively suggested to be due to activation of phosphoprotein phosphatase.[214]

III. PHOSPHATIDATE PHOSPHOHYDROLASE ACTIVITIES IN ADIPOSE TISSUES — GENERAL ASPECTS

A. Historical Development of Studies

Activity of the phosphohydrolase in rat white fat was first documented in 1960 by Rose and Shapiro[215] but it was not until 1973 that Jamdar and Fallon[19] first demonstrated that the tissue contains both Mg^{2+}-independent and Mg^{2+}-dependent forms of the enzyme. The first study in detail of the subcellular localization of phosphohydrolase activity was published in 1968 by Daniel and Rubinstein[216] but appears to have been confined to the Mg^{2+}-independent activity. Subsequent studies of the subcellular distribution of both the Mg^{2+}-independent and the Mg^{2+}-dependent activities have followed since 1973.[11,19,21] There have been few studies of the kinetic properties of either form of the enzyme from white adipose tissue (see Section III.D). Study of the control of phosphohydrolase activity in white fat has advanced in three broad directions and has almost entirely been confined to the Mg^{2+}-dependent activity. No studies of long-term changes in activity were reported until 1976 when activity changes in obesity were noted[217] followed later by studies of changes in fasting,[21,139,218] diabetes,[21] hypothyroidism,[21] age and cell size,[219-221] and in studies of adipocyte differentiation programs.[222-224] The second approach to study of the control of phosphohydrolase activity, namely the acute actions of hormones has mainly been centered in the author's laboratory, commencing in 1978, and has been concerned with the action of lipolytic hormones or insulin in vitro[10,128-131] and in vivo.[21,225] Lastly, study of the controlled translocation of phosphohydrolase activity between soluble and membraneous compartments of the cell, which is discussed elsewhere in this volume (Chaper 2, Section IV.B) actually commenced with the work of Moller et al. in 1981-82 using the adipose tissue enzyme.[20,226] It should be obvious from this short preamble that all studies in this field are quite recent.

B. Assay of Phosphatidate Phosphohydrolase Activity in Adipose Tissues

This has mainly been performed in adipose tissue extracts using an aqueous dispersion of phosphatidate as substrate. Jamdar and Fallon[19] have also investigated phosphohydrolase activity using a membrane-bound preparation of [14C] phosphatidate obtained by preincubation of adipose tissue microsomes with [U-14C]-glycerolphosphate and palmitoyl-CoA in the absence of Mg^{2+}. When rat adipose tissue, mitochondrial, microsomal, and soluble fractions were tested for phosphohydrolase activity to compare the aqueous and the membrane-bound substrates, some important differences were found.[19] With all three subcellular fractions reaction rates with the aqueous substrate were substantially higher than with the membrane-bound substrate — in fact activity is essentially unmeasurable in the mitochondrial fraction in the absence or presence of Mg^{2+} unless the aqueous dispersion is used. A similar preference for the aqueous dispersion of the substrate is also seen with liver phosphohydrolase.[227] In addition, no adipose tissue subcellular fraction has appreciable activity in the absence of Mg^{2+} when the membrane-bound substrate is used[19] whereas appreciable activity can be observed in mitochondria or microsomes in the absence of Mg^{2+} with the aqueous substrate.

Using the aqueous substrate one has the option of following the phosphohydrolase reaction either by measurement of orthophosphate release or by monitoring radioactive diacylglycerol formation from an aqueous dispersion of radiolabeled phosphatidate. Sturton and Brindley[228,229] have shown that rat liver contains phospholipase activity that deacylates phosphatidate. Subsequent dephosphorylation of the resulting glycerol phosphate then leads to overestimation of liver phosphatidate phosphohydrolase if the phosphate release assay is used, particularly with microsomal fractions. However these problems do not appear to apply to adipose tissues. Lawson et al.[225] have observed no lysophosphatidate or glycerolphosphate formation when rat liver epididymal fat pad soluble or microsomal fractions are incubated with 1,2-diacyl-[3H]-glycerol 3-phosphate. Furthermore, no dephosphorylation of [3H]-glycerolphosphate could be detected in these tissue extracts. The same applies to brown adipose tissue. Measurements of phosphatidate phosphohydrolase activity in mitochondrial, microsomal, and soluble fractions from rat interscapular brown adipocytes are essentially identical when measured either as phosphate release or diacylglycerol formation using aqueous phosphatidate as substrate.[266] Jamdar and Fallon[19] reported that rat fat pad subcellular fractions contained phosphatase activity toward glycerolphosphate. This activity had a pH optimum of 4.5 and was unaffected by Mg^{2+}. Table 3 shows that adipocyte homogenates or 100,000-g supernatants contain a very small amount of acid phosphatase activity but only a very small component of this activity was Mg^{2+}-dependent and the activity, unlike the Mg^{2+}-dependent phosphatidate phosphohydrolase activity, is very low in 100,000-g soluble fractions. It is also unaffected by exposure of the cells to noradrenaline (not shown). Likewise, Table 4 shows that even if some dephosphorylation of glycerol phosphate occurs in adipocyte extracts under conditions of the phosphohydrolase assay this is negligible compared with dephosphorylation of phosphatidate and again, unlike the dephosphorylation of phosphatidate, is unaffected by noradrenaline (not shown) and is barely affected by Mg^{2+}

In summary, phosphatidate phosphohydrolase activity in adipose tissue can be satisfactorily measured by appearance of either of the reaction products, orthophosphate or diacylglycerol, using aqueous dispersion of phosphatidate as substrate. These measurements appear to suffer little interference from other reactions.

Table 3
ACID PHOSPHATASE ACTIVITIES IN
EXTRACTS OF RAT ADIPOCYTES[268]

Extract	Acid phosphatase activity (nmol.min^{-1}.mg protein^{-1})	
	Mg^{2+}-independent	Mg^{2+}-dependent
Fat-free homogenate	18.4	1.2
105,000-g supernatant	6.2	0.2

Note: Assays were performed at pH 4.5 using *p*-nitrophenylphosphate as substrate. Mg^{2+}-dependent activities were measured using 2.5 mM Mg^{2+}.

Table 4
COMPARISON OF RATES OF HYDROLYSIS OF GLYCEROL 3-
PHOSPHATE AND PHOSPHATIDATE BY EXTRACTS OF RAT ADIPOCYTES[268]

Extract	Glycerol phosphate hydrolysis (nmol.min^{-1}.mg protein^{-1})		Phosphatidate hydrolysis (nmol.min^{-1}.mg protein^{-1})	
	Mg^{2+}-independent	Mg^{2+}-dependent	Mg^{2+}-independent	Mg^{2+}-dependent
Fat-free homogenate	2.0	4.6	4.0	54
105,000-g supernatant	1.1	1.3	2.0	45

Note: Assays were performed at pH 6.8 using 1.4 mM glycerol phosphate or phosphatidate. Mg^{2+}-dependent activities were measured using 2.5 mM Mg^{2+}.

Table 5
COMPARISON OF Mg^{2+}-INDEPENDENT AND Mg^{2+}-
DEPENDENT PHOSPHATIDATE
PHOSPHOHYDROLASE ACTIVITIES IN RAT
ADIPOSE TISSUE AND ADIPOCYTE FAT-FREE
HOMOGENATES

Source	Activity (nmol of P$_i$/min/mg protein)		Ref.
	Mg^{2+}-independent	Mg^{2+}-dependent	
Whole adipose tissue	5.4	6.9	19
	1.0	7.7	225
Adipocytes	6.8	60	128
	8.0	108	21

C. Subcellular Distribution of Phosphohydrolase Activities

Table 5 shows that in general Mg^{2+}-dependent phosphohydrolase activity predominates over the Mg^{2+}-independent activity in whole homogenates of adipose tissue or adipocytes. In addition, the specific activity of the Mg^{2+}-dependent activity is enriched by prior separation of the adipocytes from the nonadipocyte cells of the tissue. Table 6 shows that the small amount of Mg^{2+}-independent activity is distributed differently between subcellular fractions compared with the Mg^{2+}-dependent activity. In white adipocytes, Mg^{2+}-inde-

Table 6
SUBCELLULAR DISTRIBUTION OF PHOSPHATIDATE
PHOSPHOHYDROLASE ACTIVITIES IN RAT ADIPOSE TISSUE

Mg²⁺-Independent Activity

Mg^{2+}-Independent Activity

Distribution (%)

Source	Fat-free homogenate	Nuclei + mitochondria	Microsomes	Soluble fraction	Ref.
Whole tissue	100	31	16	53	216
	100	27*	42	20	19
Adipocytes	100	59	26	15	11
	100	71	29	0	21

Mg^{2+}-Dependent Activity

Whole tissue	100	1*	20	70	19
Adipocytes	100	10	21	69	11
	100	4	21	75	21

Note: All homogenates were prepared in 0.25 M sucrose medium containing 0 to 10 mM
    Tris buffer; * indicates nuclear fraction discarded.

pendent phosphohydrolase is found almost entirely in particulate fractions, predominantly in the combined nuclear and mitichondrial fractions. By contrast, Mg^{2+}-dependent phosphohydrolase is essentially absent from the nuclear and mitochondrial fractions and is predominantly a soluble enzyme in adipocytes isolated in the basal state in 0.25 M sucrose-based homogenization media. Mg^{2+}-dependent phosphohydrolase activity is also reported as being absent from adipocyte plasma membrane fractions.[226] Reference to distribution of cytosolic and microsomal marker enzymes[11,21] shows that the Mg^{2+}-dependent phosphohydrolase activity in the microsomal fraction is not attributable to cross contamination by cytosol. Hence, the adipocyte Mg^{2+}-dependent phosphatidate phosphohydrolase has a true bimodal distribution between microsomal membranes and the soluble fraction and, as considered in detail in Volume II, Chapter 6, Section VI in fact is an ambiquitous enzyme.

D. General Kinetic Properties of Mg^{2+}-Independent and Mg^{2+}-Dependent Phosphatidate Phosphohydrolases

1. K_m for Phosphatidate

From the limited number of studies (see Table 7) there is general agreement that both forms of the enzyme appear to exhibit hyperbolic kinetics with respect to the concentration of aqueous phosphatidate up to concentrations of approximately 0.5 to 1 mM. At higher concentrations a certain amount of substrate inhibition is evident[19,218] (Figure 4). All studies have found apparent K_m values between 120 and 160 μm for Mg^{2+}-dependent phosphohydrolase in fractions isolated from whole adipose tissue or from adipocytes. The limited information available suggests that the apparent K_m for phosphatidate of the Mg^{2+}-independent activity is approximately fourfold higher and is the same whether the activity is measured in mitochondria or microsomal fractions (Table 7).

2. Effect of Mg^{2+}

All studies are in agreement that Mg^{2+} concentrations of 5 mM or higher are necessary to elicit maximum or near maximum activity of the Mg^{2+}-dependent enzyme in rat adipose tissue-soluble fraction[19,218,225,230] (see also Figure 5). However, there are some unexplained

Table 7
MEASUREMENTS OF K_m FOR PHOSPHATIDATE OF PHOSPHOHYDROLASE ACTIVITIES IN RAT ADIPOSE TISSUE

Mg^{2+}-Dependent Phosphohydrolase

Source	K_m (μm)	pH of assay	$[Mg^{2+}]$ (mM)	Ref.
Adipose tissue soluble fraction	160	7.5	5	19
	160	7.5	5	230
	150	7.5	5	220
Adipocyte soluble fraction	140	6.8	20	218
	120	6.8	2.5	Figure 4

Mg^{2+}-Independent Phosphohydrolase

Adipose tissue mitochondria	550	6.8		19
Adipose tissue microsomes	550	6.8		19

Note: All measurements were made using aqueous phosphatidate as substrate.

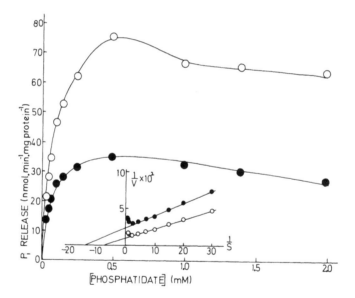

FIGURE 4. Mg^{2+}-dependent phosphatidate phosphohydrolase in 105,000-g supernatants from rat adipocytes — effect of phosphatidate concentration. Adipocytes were incubated for 30 min in media containing 5 mM glucose and albumin (40 mg/mℓ) with (●) or without (○) 0.5 μM noradrenaline before homogenization and preparation of soluble fractions. Mg^{2+} concentration in the assays was 2.5 mM. K_m values determined from the inset. Lineweaver-Burke plots were 70 and 120 μM in the presence and absence of noradrenaline, respectively. (The data are unpublished work by the author.)

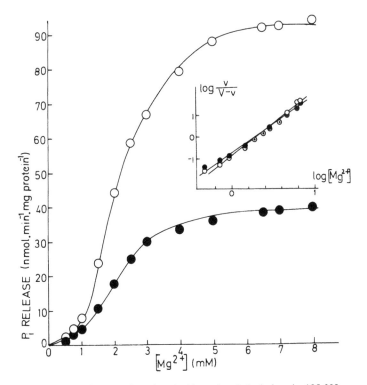

FIGURE 5. Mg^{2+}-dependent phosphatidate phosphohydrolase in 105,000-*g* supernatants from rat adipocytes — effect of Mg^{2+} concentration. Cells were treated with (●) or without (○) 0.5 μ*M* noradrenaline for 10 min as in Figure 4. Phosphatidate concentration in the assays was 1.4 m*M*. $K_{0.5}$ value determined from the inset Hill plots was 2 m*M* in both cases. Hill coefficients were 2.6 and 3.0 with and without noradrenaline treatment. (The data are unpublished work by the author.)

differences between different studies. Jamdar and Fallon[19] observed some inhibition of activity at Mg^{2+} concentrations in excess of 5 m*M* whereas this effect has not been noted subsequently[218,230] (Figure 5). Moller et al.,[218] Lawson et al.,[225] and Jamdar and Osborne[230] have observed essentially hyperbolic relationships between rat adipose tissue cytosolic phosphohydrolase activity and the concentration of Mg^{2+} whereas Jamdar and Fallon[19] found this relationship to be sigmodial, a phenomenon also observed by the author (see Figure 5). The degree of divergence from hyperbolic kinetics with respect to Mg^{2+} concentration shown in Figure 5 is not attributable to the small amount of EDTA (0.1 m*M*) that is incorporated into the assay when added with the tissue extract. The reasons for these discrepancies between different studies are unclear at present but it may be noteworthy that hyperbolic kinetics with respect to Mg^{2+} concentration have been observed in assays containing albumin[218,230] or phosphatidylcholine.[225]

When rat adipose tissue microsomal fractions are tested for Mg^{2+} dependence the activity vs. [Mg^{2+}]-curves are shifted to the left compared with those seen with soluble fraction such that 1 to 2 m*M* Mg^{2+} generally elicits maximum phosphohydrolase activity.[19,225] Furthermore, the microsomal activity shows considerable inhibition when the Mg^{2+} is increased to higher concentrations.

3. Effects of Other Divalent Cations

Jamdar and Fallon[19] have tested the effects of Co^{2+}, Ni^{2+}, Zn^{2+}, Cu^{2+}, and Mn^{2+} on adipose tissue phosphohydrolase activity in microsomes and soluble fraction. Using membrane-bound substrate which already contained sufficient Mg^{2+} to activate the enzyme, all of these other divalent cations were inhibitory. When aqueous phosphatidate was used as substrate none of these anions could mimic the stimulatory effect of Mg^{2+}; and Zn^{2+}, Co^{2+}, Ni^{2+}, and Mn^{2+} at pH 7.5 actually antagonized the effect of Mg^{2+}. Ca^{2+} is also reported to inhibit Mg^{2+}-dependent and Mg^{2+}-independent activities.[216,225]

4. Effect of pH

Phosphohydrolase activity in whole homogenates from rat tissue assayed in the absence of Mg^{2+} was reported as being optimal at pH 6.[216] Jamdar and Fallon[19] found the optimum pH of the Mg^{2+}-independent enzyme to be 6.8 in both mitochondrial and microsomal fractions. The pH optimum of the Mg^{2+}-dependent activity does not appear to have been measured directly at physiological concentrations of $Mg.^{2+}$ The degree of activation by Mg^{2+} appears to vary slightly with pH.[19,218]

5. Effect of Thiol-Group Reagents

Mg^{2+}-dependent phosphohydrolase in adipose tissue microsomes or soluble fraction is considerably inhibited by N-methyl- or N-ethylmaleimide with 5 mM concentrations of these agents being sufficient to decrease activity by approximately 80% in a manner that is not competitive with respect to phosphatidate concentration.[220] By contrast, the Mg^{2+}-independent activity in microsomes or mitochondria is unaffected or even slightly stimulated by these agents.[220] This differential sensitivity to these reagents would appear to offer a simple and convenient way of differentiating between the two forms of the enzyme.

6. Thermolability

Mg^{2+}-dependent phosphohydrolase activity in adipose tissue microsomes or soluble fraction shows considerable thermolability with approximately 90% of the activity being lost on incubation at 55°C for 10 min.[220] By contrast, the Mg^{2+}-independent activity in mitochondria or microsomes is completely stable to this treatment.[220]

7. Susceptibility to Proteolytic Enzymes

Adipose tissue Mg^{2+}-dependent phosphohydrolase in microsomes and soluble fraction is considerably more susceptible to inactivation by trypsin, α-chymotrypsin, protease, and proteinase K than is the Mg^{2+}-independent activity in microsomes and mitochondria.[220] This differential susceptibility is seen whether or not membranes are disrupted by detergent.

8. Effect of Polyamines

Spermine, spermidine, and putrescine all activate Mg^{2+}-dependent phosphatidate phosphohydrolase in rat adipose tissue microsomes or soluble fraction.[230] On the other hand, spermine and spermidine are inhibitory towards the Mg^{2+}-independent activity in microsomes while putrescine has little effect on this activity.[230] In the absence of Mg^{2+}, polyamines have little effect, i.e., these polyvalent cations do not replace Mg^{2+} but appear to potentiate its stimulatory effect on the Mg^{2+}-dependent phosphohydrolase.[230] The polyamines increase the V_{max} and slightly decrease the K_m for phosphatidate from 160 to 100 μm for the cytosolic Mg^{2+}-dependent enzyme.[230] The stimulatory effect of polyamines was observed both in the presence and the absence of the activator phosphatidylcholine (see Section III.D.9) suggesting that these two modulators may act through different mechanisms.[230]

9. Effects of Glycerolipids

Phosphatidylcholine activates both Mg^{2+}-dependent and Mg^{2+}-independent phosphohydrolases in adipose tissue[218,225,230] causing as much as a doubling in activity. By contrast, phosphatidylethanolamine, phosphatidylinositol, and diacylglycerol are inhibitory and phosphatidylserine has little effect.[218]

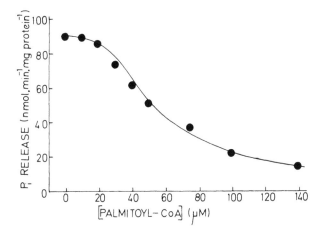

FIGURE 6. Mg^{2+}-dependent phosphatidate phosphohydrolase in 105,000-*g* supernatant from rat adipocytes — inhibition by palmitoyl-CoA. Assays were performed using 1.4 m*M* phosphatidate and 1.5 m*M* Mg^{2+}. (The data are unpublished work by the author.)

10. Effects of Fatty Acids and Fatty Acyl-CoA Thioesters

Moller et al.[218] have observed inhibition of Mg^{2+}-dependent phosphohydrolase in the rat adipocyte cytosolic fraction by oleate and by oleoyl-CoA. Oleate (0.8 m*M*) was necessary to elicit 50% inhibition. Since these assays contained albumin at 0.4 mg/mℓ, this concentration would represent a fatty acid/albumin molar ratio in the assay of 132; i.e., quite high unbound concentrations of oleate would be present and it seems rather unlikely that this reported effect of oleate has any direct physiological relevance. Oleoyl-CoA was found to be a more potent inhibitor than oleate by Moller et al.[218] The concentration-dependence of inhibition of the rat cytosolic Mg^{2+}-dependent enzyme by palmitoyl-CoA is shown in Figure 6. These assays contained 15 μ*M* albumin (1 mg/mℓ) and 50% inhibition was elicited by 58 μ*M* palmitoyl-CoA (an acyl-CoA/albumin molar ratio of 3.8). It is therefore possible that fatty acyl-CoA thioesters might have some role in the physiological regulation of the Mg^{2+}-dependent phosphohydrolase (but see also Chapter 2, Section IV.B).

11. Effect of Albumin

Inclusion of albumin in assays decreases the activity of the Mg^{2+}-dependent phosphohydrolase from rat adipocyte soluble fractions — presumably due to hydrophobic interaction between the albumin and the aqueous dispersion of the substrate. Figure 7 shows the close agreement between two separate studies of this effect; 50% inhibition of activity is seen with approximately 230 μ*M* albumin (15 mg/mℓ).

E. Comparison of Phosphatidate Phosphohydrolase Activities in Adipose Tissue with Those in Other Tissues

It is not easy to make such comparisons unless measurements are made in the same laboratory under comparable conditions. Jamdar et al.[220] have made measurements in liver, kidney, brain, heart, and white adipose tissue of 3-month-old rats. They report that microsomal Mg^{2+}-independent and Mg^{2+}-dependent activities from whole adipose tissue are at least double those in any of these other tissues. Their reported values for the microsomal Mg^{2+}-dependent activity in adipose tissue and liver were approximately 10 and 12 nmol.min^{-1}.mg protein^{-1} respectively. The very high activity of adipose tissue Mg^{2+}-dependent activity compared with other tissues is even more evident in soluble fractions where Jamdar et al.[220] obtained activities of approximately 14 nmol.min^{-1}.mg protein^{-1}

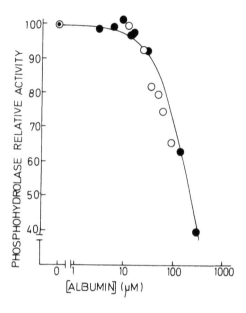

FIGURE 7. Mg^{2+}-dependent phosphatidate phosphohydrolase activity in rat adipocyte soluble fractions — decrease in activity with albumin. (●) Assayed at pH 6.8 with 1.4 mM phosphatidate and 2.5 mM Mg^{2+}.[268] (○) Assayed at pH 7.5 with 1.14 mM phosphatidate and 5 mM Mg^{2+}.[218]

whereas this activity was only approximately 1 to 2 nmol.min^{-1}.mg protein^{-1} in cytosolic fraction from liver, kidney, brain, and heart. By contrast, all five tested tissues, including adipose tissue, had activities for the Mg^{2+}-independent phosphohydrolase in the cytosolic fraction of approximately 2 nmol.min^{-1}.mg protein^{-1}. Reasonably direct comparison may also be made from the studies of Sturton and Brindley[229] and Lawson et al.[225] which show maximal rates of Mg^{2+}-dependent phosphohydrolase in soluble fractions of liver and adipose tissue to be approximately 2 and 22 nmol.min^{-1}.mg protein^{-1}, respectively.

These studies actually substantially underestimate the activity for the Mg^{2+}-dependent phosphohydrolase in adipose tissue since isolation of subcellular fractions from whole adipose tissue incorporates material from nonadipocytes. The adipocytes are the active component of the tissue with regard to triacylglycerol synthesis. Table 8 shows collected values for the Mg^{2+}-dependent activity in soluble fractions prepared either from whole tissue or after isolation of the adipocytes. As far as can be ascertained these values were obtained under near-optimal assay conditions and suggest that soluble Mg^{2+}-dependent phosphohydrolase activity is enriched six- to sevenfold on separation of the adipocytes. At present there are no published values for phosphohydrolase activities in brown fat but preliminary study in the author's laboratory show that soluble fractions from interscapular brown adipocytes contain only approximately 20% of the Mg^{2+}-dependent phosphohydrolase specific activity that is found in similar fractions of white adipocytes obtained from the same animals.[266]

IV. SHORT-TERM REGULATION OF PHOSPHATIDATE PHOSPHOHYDROLASE ACTIVITY

A. Changes Observed In Vitro
1. Effects of Noradrenaline and Insulin
When rat white adipocytes are incubated for short times with lipolytic agents relatively

Table 8
COMPARISON OF Mg²⁺-DEPENDENT
PHOSPHATIDATE
PHOSPHOHYDROLASE ACTIVITIES IN
SOLUBLE FRACTIONS OBTAINED
EITHER FROM WHOLE ADIPOSE
TISSUE OR FROM ADIPOCYTES
PREVIOUSLY OBTAINED FROM SUCH
TISSUES

Source	Activity (nmol of P_i/min/mg of protein)	Ref.
Rat whole tissue	10	19
	22	225
	18	230
	14	220
Mouse whole tissue	22	217
Rat adipocytes	132	218
	100	20
	91	226
	94	Figure 5

Note: All studies used an aqueous dispersion of phosphatidate as substrate.

persistent modifications in Mg²⁺-dependent phosphatidate phosphohydrolase activity can be observed under appropriate conditions.[10,128-131] The experimental procedure involves rapid recovery of the cells from the incubation medium (approximately 20 sec) followed by freeze-stopping in liquid nitrogen. Following this each cell sample is individually homogenized, briefly centrifuged to remove the bulk fat, and immediately assayed for phosphohydrolase activity. This is generally commenced within 1.5 to 2 min of the homogenization of the samples.

This sort of procedure has been widely used in metabolic studies to detect enzyme activity changes due to covalent modification since the freeze-stopping and homogenization procedures respectively arrest and, by dilution, slow down the actions of interconversion enzymes such as protein kinases and phosphoprotein phosphatases. By such a procedure enzymes subject to control by covalent modification can usually be retained in an activity state approaching that within the intact cell provided assays rapidly follow homogenization. It is also possible that such a procedure may also detect noncovalent modification of an enzyme's activity provided the interaction between the enzyme and regulatory ligand(s) is tight; i.e., the rate of their dissociation is slow. By use of this freeze-stop/rapid assay procedure Cheng and Saggerson[128-130] have shown a decrease in rat adipocyte Mg²⁺-dependent phosphohydrolase activity on brief exposure of the cells to noradrenaline (Figure 8). This is part of a more general set of phenomena whereby other enzyme activities in the adipocyte triacylglycerol synthesis pathway are also rapidly decreased by the catecholamine hormones. These other activities are glycerolphosphate acyltransferase,[12,124,125] dihydroxyacetonephosphate acyltransferase,[14] acyl-CoA synthetase,[126,127] and diacylglycerol acyltransferase[132] (see Section II.A.2.b). By contrast, catecholamine-induced decreases in activity are not seen with the Mg²⁺-independent phosphohydrolase[128] (see Figure 9), phosphatidate cytidylyl transferase,[131] and CDP choline phosphotransferase[131] suggesting a selectivity of this mode of control for the triacylglycerol synthesis pathway. With 0.5 μ*M*

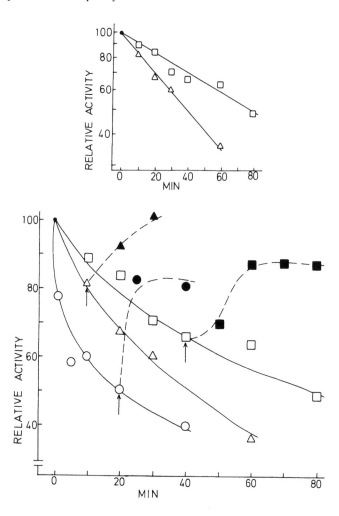

FIGURE 8. Rapid effects of noradrenaline and insulin on activities of triglyceride synthesizing enzymes in rat adipocytes. All cell incubations were in media containing 5 mM glucose and albumin (40 mg/mℓ). The data are derived from Hall and Saggerson[127] for acyl-CoA synthetase ($\bigcirc$,$\bullet$), from Rider and Saggerson[12] for glycerolphosphate acyltransferase ($\square$,$\blacksquare$) and Cheng and Saggerson[129] for Mg^{2+}-phosphatidate phosphohydrolase ($\triangle$,$\blacktriangle$) the activities of which were measured in defatted homogenates from freeze-stopped cells. Incubations were commenced with 0.5 or 1 μM noradrenaline alone present (open symbols), and at the times indicated by the arrows insulin (0.2 or 0.4 munit/mℓ) was added (closed symbols). The inset figure shows that phosphohydrolase and glycerolphosphate acyltransferase activities decline exponentially with time in the presence of noradrenaline with halftimes of 40 and 80 min, respectively.

noradrenaline, which is a maximally effective concentration,[128] Mg^{2+}-dependent phosphohydrolase activity declines exponentially with time (Figure 8) such that 50% of the starting activity is lost in approximately 40 min. This may be contrasted with half-times of inactivation for acyl-CoA synthetase and glycerolphosphate acyltransferase of 20 min and 80 min, respectively (Figure 8). A further feature of these activity modifications is that they are all rapidly reversible within cells on subsequent addition of insulin (Figure 8). These effects of noradrenaline and insulin on Mg^{2+}-dependent phosphohydrolase activity are observed in the absence or the presence of glucose in cell incubation media[128,129] suggesting that they cannot

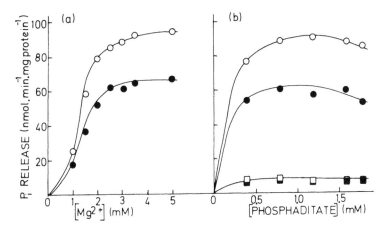

FIGURE 9. Effect of noradrenaline treatment of rat adipocytes on kinetics of Mg^{2+}-dependent and Mg^{2+}-independent phosphatidate phosphohydrolases. Cells were incubated in media containing 5 mM glucose and albumin (40 mg/mℓ) for 30 min with (closed symbols) or without (open symbols) noradrenaline (0.5 μM) followed by freeze-stopping, preparation of defatted homogenates, and then rapid assay. ($\bigcirc$,$\bullet$) Mg^{2+}-dependent phosphohydrolase, ($\square$,$\blacksquare$) Mg^{2+}-independent phosphohydrolase. (a) Assays were performed with 1.4 mM phosphatidate; (b) assays were performed with or without 2.5 mM Mg^{2+}.[268]

Table 9
CHANGES IN Mg²⁺-DEPENDENT PHOSPHATIDATE PHOSPHOHYDROLASE ACTIVITY ON INCUBATION OF RAT ADIPOCYTES WITH VARIOUS AGENTS[129,131]

Incubation time (min)	Additions	Mg²⁺-dependent phosphohydrolase activity relative to that in untreated cells	Glycerol accumulation in incubation media (mM)
60	None	100	0.03
	Noradrenaline (0.1 μM)	63	0.73
	Noradrenaline (0.1 μM) + insulin (0.2 munit/mℓ)	100	0.05
	Insulin (0.2 munit/mℓ)	97	0.04
	Dibutyryl cAMP (2 mM)	64	0.72
	Dibutyryl cAMP (2 mM) + insulin (0.2 munit/mℓ)	63	0.93
20	None	100	0.02
	Noradrenaline (0.5 μM)	59	0.76
	Propranolol (2.5 μM)	101	0.02
	Noradrenaline (0.5 μM) + propranolol (2.5 μM)	92	0.18
30	None	100	—
	Palmitate (2 mM)	88	—
	Palmitate (4 mM)	80	—

be secondary to any effects of these hormones upon membrane transport of the sugar (see Section II.A.2.a.ii). Addition of insulin at the start of the incubations, as expected, blocks the effect of noradrenaline (Table 9). However insulin alone has no effect on the phosphohydrolase activity. Figure 9 shows that the inactivating effect of noradrenaline is a "V_{max}" effect with respect to Mg^{2+} and phosphatidate and also emphasizes that the small Mg^{2+}-independent phosphohydrolase activity is not affected by the hormone. In many ways

Table 10
EC$_{50}$ VALUES FOR INACTIVATION OF TRIACYLGLYCEROL-SYNTHESIZING ENZYMES AND STIMULATION OF LIPOLYSIS ON INCUBATION OF RAT ADIPOCYTES WITH CATECHOLAMINE HORMONES

Enzyme	Agonist	EC$_{50}$ for enzyme inactivation (nM)	EC$_{50}$ for stimulation of lipolysis	Ref.
Mg^{2+}-dependent phosphohydrolase	Noradrenaline	90	$0.7 \times 10^{-7} M$	128
Acyl-CoA synthetase	Noradrenaline	100	$1.0 \times 10^{-7} M$	127
Glycerolphosphate acyltransferase	Adrenaline	200	$3.0 \times 10^{-7} M$	124

Note: The values are calculated from data in the indicated studies in which both parameters were measured simultaneously.

this inactivation of the phosphohydrolase (and the other four enzymes named above) would appear to be closely related to the simultaneous stimulation of lipolysis. The physiological rationale for this has been described in Section II.A.2.b. Consideration of the EC$_{50}$ values for inactivation of triglyceride-synthesizing enzymes and stimulation of lipolysis by catecholamines (Table 10) shows a very close parallelism between the two processes. Further parallels are the blockade of both processes by the β-adrenoceptor antagonist propranolol and the effect of dibutyryl cAMP on both processes (Table 9). One therefore has to consider the possibility that this relatively persistent modification of the phosphohydrolase activity might share a common mechanism with the control of the hormone-sensitive lipase; i.e., covalent modification by a phosphorylation/dephosphorylation mechanism (see Section II.B). An alternative possibility is that the inactivation of the phosphohydrolase might be secondary to changes in the concentration of regulatory substances brought about by the process of lipolysis. The first possibility can be eliminated as a direct means for control of the enzyme activity by a relatively simple experiment. Normally, after freeze stopping, adipocytes are homogenized in a medium consisting of 0.25 M sucrose containing 10 mM Tris buffer (pH 7.4), 1 mM EDTA, and 1 mM dithiothreitol.[128] If defatted albumin (10 mg/mℓ) is also incorporated into this homogenization medium the prior effect of noradrenaline is abolished in that the Mg^{2+}-dependent phosphohydrolase activity is restored to the control level. On addition of these homogenates to the phosphohydrolase assay mixture a tenfold dilution occurs such that albumin concentrations of as little as 15 μM in the assay are sufficient to abolish the noradrenaline effect (Figure 10). Such a concentration of albumin has little effect upon the phosphohydrolase activity in extracts from untreated cells (Figure 10; also see Figure 7). A similar abolition of the inactivating effect of noradrenaline by albumin is also seen in the case of the microsomal glycerolphosphate acyltransferase[12] but not in the case of the acyl-CoA synthetase.[127] Simple reversal by albumin strongly implies that the effect of noradrenaline cannot be attributed to a covalent modification and suggests that the Mg^{2+}-dependent phosphohydrolase must be inactivated by a ligand (or ligands) that binds to albumin. Since the effect persists into homogenates the interaction with the phosphohydrolase must be a tight one. There is an additional possibility that cannot at present be discounted, namely that the phosphohydrolase might actually undergo a covalent modification which by itself may not alter the enzyme activity but yet changes its sensitivity to an inhibitory ligand. Attempts to cause changes in the activity of the Mg^{2+}-dependent phosphohydrolase by incubation of rat adipocyte soluble fraction with cAMP-dependent protein kinase or with protein kinase C have been unsuccessful.[267] A similar lack of effect of cAMP-dependent protein kinase on the hepatic phosphohydrolase has also been reported.[231]

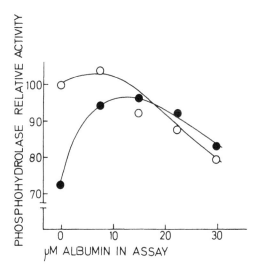

FIGURE 10.   Abolition of the effect of noradrenaline on rat adipocyte Mg^{2+}-dependent phosphatidate phosphohydrolase by inclusion of albumin in homogenization media. Cells were incubated for 30 min in media containing 5 mM glucose and albumin (40 mg/mℓ) with (●) or without (○) noradrenaline (0.5 μM) followed by freeze-stopping preparation of defatted homogenates in media containing 0 to 300 μM albumin and then rapid assay of 0.05-mℓ aliquots of the homogenates in a final assay volume of 0.5 mℓ.

Since nonesterified fatty acids are products of lipolysis (see Chapter 6, Section II.B) and are known to bind to albumin it might be concluded that these represent the sole regulatory ligand(s) generated when the cells are treated with noradrenaline or dibutyryl cAMP. Various arguments can be advanced against this proposal. First, although incubation of adipocytes with high concentrations of palmitate causes a small inactivation of the phosphohydrolase[129] (Table 9) this change is small compared with the degree of inactivation caused by concentrations of noradrenaline or dibutyryl cAMP which bring about similar or smaller nonesterified fatty acid levels.[129] In a general sense the same conclusion is reached by consideration of Figure 11. This is a summary of a number of different experiments and shows that phosphohydrolase activity and nonesterified fatty acid accumulation show a close inverse relationship under a variety of conditions. This is not surprising in view of the time dependence of the phosphohydrolase inactivation (Figure 8), the time dependence of nonesterified fatty acid accumulation, and the similarity in the noradrenaline dose relationships for these two processes (Table 10). What is also evident though is that this relationship between phosphohydrolase activity and fatty acid concentration is less steep when fatty acid (palmitate) is simply added in the absence of a lipolytic agent. Thus, although fatty acid may cause some inactivation of the phosphohydrolase directly (or indirectly through provision of acyl-CoA derivatives, see Section III.D.10), factors other than fatty acid alone must contribute to the effect of the lipolytic agents. A second argument is that the concentration of nonesterified fatty acid likely to be encountered in the defatted homogenates used in these studies is not more than 0.6 mM[232,233] which, when diluted 1:10 into the phosphohydrolase assay gives a final level of only approximately 60 μM which is not an appreciable inhibitory concentration for the enzyme (see Moller et al.[218] and Section III.D.10). Third, when insulin is used to rapidly reverse the effect of noradrenaline (see Figure 8) complete reactivation of the enzyme can be achieved while fatty acid accumulation remains.[129]

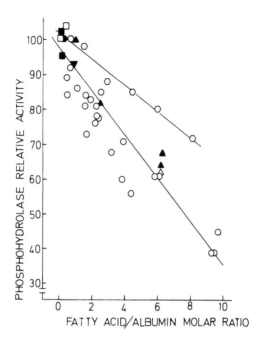

FIGURE 11. Relationship between Mg^{2+}-dependent phosphohydrolase activity in extracts of freeze-stopped adipocytes and the concentration of nonesterified fatty acid in the incubation media from which they were recovered. The data are derived from Cheng and Saggerson[128,129] and Cheng et al.[131] Phosphohydrolase activity and extracellular fatty acid concentration were perturbed by incubation either with various concentrations of lipolytic agents for various times or by incubation with palmitate. The fatty acid concentrations are those at the termination of the incubations. All incubations contained 5 mM glucose. (●) Control; (○) with noradrenaline; (■) with insulin; (□) with noradrenaline + insulin; (▼) with noradrenaline + propranolol; (▲) with dibutyryl cAMP; (△) with dibutryl cAMP + insulin; (○) with palmitate. Regression lines are (1) through the data for incubation with palmitate and for control incubations: r = −0.991 and slope = −3.73 (n = 7). (2) through the data for incubation with lipolytic agents and for control incubations: r = −0.939 and slope = −6.13 (n = 41).

Figure 12 illustrates that noradrenaline inactivation is a stable phenomenon provided whole defatted homogenates from adipocytes are maintained at 0 to 4°C. However, if homogenates from untreated cells are maintained at 37°C for a few minutes the Mg^{2+}-dependent phosphohydrolase activity rapidly declines. This trend is far less pronounced with homogenates from noradrenaline-treated cells with the result that incubation of untreated cell homogenates at 37°C (but not 0°C or 20°C)[130] in part mimics the effect of noradrenaline. By contrast, Mg^{2+}-dependent phosphohydrolase activity in 100,000-g soluble fractions from untreated cells is stable at 37°C. One possible interpretation is that a regulatory factor is generated by a particulate component of the homogenate at 37°C and that this may represent the regulatory ligand that is generated by noradrenaline treatment of the cells. The regulatory ligand (or ligands) remains unidentified at present.

Figure 13 shows that mixtures of homogenates from noradrenaline-treated and untreated

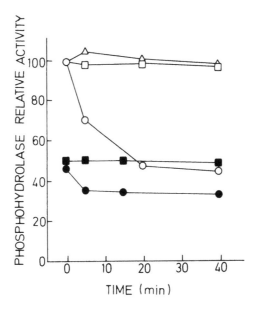

FIGURE 12. Stability of Mg^{2+}-dependent phosphohydrolase in adipocyte extracts on incubation at 0 and 37°C. Cells were incubated for 30 min with (closed symbols) or without (open symbols) noradrenaline (0.5 μM) in media containing 5 mM glucose and albumin (40 mg/mℓ). The cells were recovered by centrifugation and either freeze-stopped and homogenized to yield defatted whole homogenates or were gently broken without freezing and 105,000-g supernatants isolated. Phosphohydrolase activities are expressed relative to that in the appropriate control extract. The data are derived from Saggerson et al.[10] and Cheng and Saggerson.[130] (○,●) Whole homogenates incubated at 37°C; (□,■) whole homogenates incubated at 0°C; (△) 105,000-g supernatant from untreated cells incubated at 37°C.

cells yield phosphohydrolase activities in subsequent assays that are the strict arithmetic sum of the starting activities suggesting that there is little excess or unbound regulatory ligand in the extracts from the noradrenaline-treated cells. Finally, dialysis of homogenates from noradrenaline-treated and untreated cells does not alter phosphohydrolase activity in either homogenate and does not abolish the effect of noradrenaline.[268]

2. Effects of Phorbol Esters

The phorbol ester tumor promoter 12-*O*-tetradecanoylphorbol 13-acetate (TPA) mimics certain actions of insulin in rat adipocytes by increasing 3-*O*-methyl glucose transport[234] and glucose metabolism.[235,236] At the same time, rather perversely, TPA antagonizes the stimulation of these processes by insulin.[234,236] The most clearly recognized action of TPA and related compounds is their stimulation of protein kinase C by substituting for diacylglycerol which is required to activate the enzyme.[237,238] Protein kinase C is reported to be present in rat adipose tissue[239-241] and TPA appears to promote a translocation of protein kinase C from the soluble compartment to particulate components within adipocytes.[235] Since TPA can also mimic some of the effects of vasopressin in hepatocytes[242] and vasopressin causes activation of hepatocyte phosphatidate phosphohydrolase,[243] the effect of TPA was tested on Mg^{2+}-dependent phosphohydrolase in rat adipocytes.[244] Figure 14 shows that the phosphohydrolase

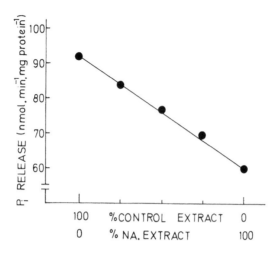

FIGURE 13.   The result of mixing defatted homogenates from noradrenaline-treated and untreated adipocytes. Rat adipocytes were incubated for 30 min in media containing 5 mM glucose and albumin (40 mg/mℓ) with or without noradrenaline (0.5 μM) followed by freeze-stopping and production of defatted homogenates. The extracts from treated and untreated cells were then mixed in various proportions and incubated at 0°C for 30 min before being assayed for Mg^{2+}-dependent phosphohydrolase activity. (●) Represents the measured activities and the line indicates the expected values if these were the arithmetic sum of the two mixed components.[268]

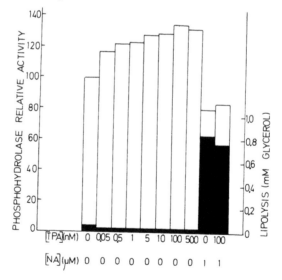

FIGURE 14.   Effect of 12-*O*-tetradecanoylphorbol 13-acetate on adipocyte Mg^{2+}-dependent phosphohydrolase activity and lipolysis. Rat adipocytes were incubated for 30 min in media containing 5 mM glucose and albumin (40 mg/mℓ) together with the indicated concentration of phorbol ester (TPA) or noradrenaline (NA). The open histogram represents phosphohydrolase activity relative to the control and the filled part of the histogram represents glycerol accumulation in the incubation media.[244]

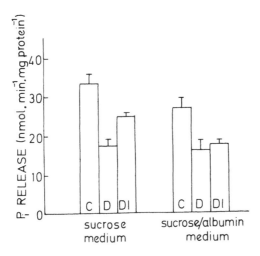

FIGURE 15. Rapid effect of insulin on Mg^{2+}-dependent phosphohydrolase activity in adipose tissue from streptozotocin-diabetic rats. The data were derived from Taylor and Saggerson[21] and represent means ± SEM from 6 to 7 separate animals where epididymal adipose tissues were either homogenized in 0.25 M sucrose containing 10 mM Tris-HCl (pH 7.4), 1 mM EGTA and 1 mM dithiothreitol (sucrose medium), or in this sucrose medium containing 10 mg/mℓ albumin (sucrose-albumin medium). Activities were measured in defatted whole homogenates. C = Control; D = streptozotocin diabetic (100 mg/kg, 2 days previously); DI = diabetic 2 hr after insulin injection (20 units/kg).

activity in defatted homogenates was increased in a dose-dependent manner when adipocytes were exposed to TPA for 30 min. The EC_{50} for this effect was approximately 100 pM TPA which may be contrasted with EC_{50} values for effects of TPA on glucose transport and metabolism varying between 1 pM[234] to 10 nM.[235,236] Noradrenaline overrode the stimulatory effect of TPA (Figure 14) and it is noteworthy that TPA had no effect on basal lipolysis and did not mimic the antilipolytic action of insulin (Figure 14). The mechanisms underlying this effect of the phorbol ester on the phosphohydrolase remain obscure at present.

B. Changes In Vivo
1. Effect of Corticortopin
Lawson et al.[225] injected rats with corticotropin (1 unit/kg = 10 μg/kg body weight) and observed 20 to 25% decreases in white adipose tissue Mg^{2+}-dependent phosphatidate phosphohydrolase activity between 30 and 100 min thereafter. This effect was reproducibly observed between November and December with rats fed either high carbohydrate or high fat diets.[245] However, the decreases were not seen from March to September in rats that were maintained on a standard 41B rat diet.[225]

2. Effect of Insulin
Although insulin alone is ineffective against phosphohydrolase activity when added to incubations of cells from normal animals (see Section IV.A.1) Taylor and Saggerson[21] have found that Mg^{2+}-dependent phosphatidate phosphohydrolase activity in adipose tissue homogenates is increased by 44% within 2 hr if streptozotocin-diabetic rats are injected with insulin at 20 units/kg, i.e., 1 mg/kg body weight (Figure 15). It is most likely that this

Table 11
CHANGES IN RAT ADIPOSE TISSUE Mg^{2+}-DEPENDENT
PHOSPHATIDATE PHOSPHOHYDROLASE ACTIVITY IN
STARVATION

Source	Days of starvation	% Activity in the fed state		Ref.
		Relative to protein	Relative to DNA or cell number	
Adipocyte soluble fraction	24	83	63	218
	48	85	52	
	72	76	36	
Adipose tissue homogenate	24	113	—	225
	48	96	—	
Adipocyte homogenate	24	95	104	139
	48	93	62	
	72	68	55	
Adipose tissue homogenate	48	50	—	21

change represents an acute effect of the hormone rather than a change in the amount of the phosphohydrolase protein since inclusion of albumin (150 μM) in homogenization buffers reduced the insulin-induced increase to only 10%. It is suggested that albumin may bind to some stimulatory ligand which may be a mediator of this effect of insulin. There are obvious parallels between this effect of insulin in vivo and the effect of insulin in vitro to reactivate the enzyme after noradrenaline treatment (see Section IV.A.1).

V. LONG-TERM REGULATION OF PHOSPHATIDATE PHOSPHOHYDROLASE ACTIVITY

There is a growing body of knowledge on the effects of changing nutritional or hormonal status over periods of days on the activity of the rat adipose tissue Mg^{2+}-dependent phosphohydrolase. A striking feature that emerges is the considerable divergence between adipose tissue and liver in the response of the enzyme to some of these states. All studies at present are limited by the lack of specific antisera to the enzyme so that it is not possible to establish clearly the extent to which recorded changes in enzyme activity reflect alterations in the amount of enzyme protein rather than modification of existing enzyme.

A. Fasting

Table 11 shows that there generally is some decrease in the adipose tissue Mg^{2+}-dependent phosphohydrolase activity over a period of fasting although the size of this effect varies between laboratories. The effect appears to be more pronounced when expressed relative to DNA or cell number rather than relative to tissue protein content. Refeeding 72- or 96-hr starved rats for 48 hr restores the phosphohydrolase activity to the control value.[139] By contrast, hepatic Mg^{2+}-dependent phosphohydrolase is increased by fasting.[246]

B. Diabetes

Streptozotocin-diabetes results in a 49% decrease in Mg^{2+}-dependent phosphohydrolase activity in rat adipose tissue homogenates relative to the tissue protein content[21] (see Figure 15) and 2 days of insulin replacement (20 units/kg body weight daily for 2 days) restores the activity to the control level.[21] This longer-term restoration of activity differs from the acute effect of insulin described in Section IV.B.2 in that it is observed with or without the inclusion of albumin in homogenization buffers.[21] Again, by contrast, diabetes causes the hepatic phosphohydrolase activity to increase[247-249] and insulin acts, in opposition to glucagon or glucocorticoids (Chapter 2, Section II.A.2), to decrease hepatocyte phosphohydrolase activity.[250-253]

C. Thyroid Status

Hypothyroidism caused by feeding a low iodine diet and propylthiouracil decreases rat adipose tissue Mg^{2+}-dependent phosphohydrolase activity relative to tissue protein by 79%.[21] Three days of thyroxine replacement therapy restores the activity by 2.6-fold back to 61% of the euthyroid value.[21]

D. Ethanol Feeding

This procedure causes no change in adipose tissue phosphohydrolase activity[254] but dramatically increases the activity of the hepatic enzyme (Chapter 2, Section II.B).

E. Changes in Dietary Composition

Dietary modification by enriching rat diets with starch, sucrose, corn-oil, or tallow has no significant effect on the activity of the adipose tissue Mg^{2+}-dependent or Mg^{2+}-independent phosphohydrolase activity.[225]

F. Age and Cell Size

Homogenates of larger adipocytes are more active in glycerolipid synthesis than those of smaller cells and this difference is correlated with parallel changes in the activity of Mg^{2+}-dependent phosphohydrolase[219] and also of other glycerolipid synthesizing enzymes. In rat adipose tissue homogenates glycerolipid synthesis reaches a maximum rate at approximately 60 days of age and thereafter declines despite later increases in adipocyte size.[255] This is paralleled by decreases in the tissue activity of Mg^{2+}-dependent phosphohydrolase but not of Mg^{2+}-independent activity.[220] An age-dependent decrease in the Mg^{2+}-dependent phosphohydrolase activity is observed in both larger and smaller adipocytes.[221]

G. Obesity

Adipose tissue homogenates and microsomal fractions from the obese (ob/ob) mouse strain exhibit elevated rates of glycerolipid synthesis compared with those from lean animals and this is paralleled by substantial increases in both Mg^{2+}-independent and Mg^{2+}-dependent phosphohydrolase activities in the microsomes and the cytosolic fraction.[217] There has been one report[256] that phosphohydrolase activity is increased in adipose tissue of obese humans compared with control subjects but unfortunately no experimental data were provided.

H. Adipocyte Differentiation

Phosphatidate phosphohydrolase activity has been measured during the course of the differentiation of 3T3-L1 preadipocytes to adipocytes[222,223] and has not been found to increase as substantially as the activity of other lipogenic enzymes. However, it is unclear whether the Mg^{2+}-dependent or the Mg^{2+}-independent activity was measured in these studies.

VI. REGULATION OF THE TRANSLOCATION OF Mg^{2+}-DEPENDENT PHOSPHATIDATE PHOSPHOHYDROLASE BETWEEN CYTOSOL AND MEMBRANES

A. Studies in Cell-Free Systems

The description of the subcellular distribution of the enzyme in Section III.C refers to studies in which tissues or cells were disrupted in homogenization media of low ionic strength

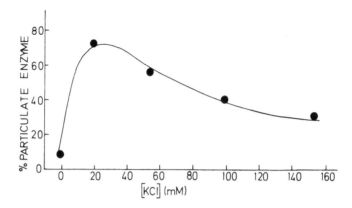

FIGURE 16.   Effects of KCl in homogenization media on the proportion of Mg^{2+}-dependent phosphohydrolase that is found in the microsomal fraction. The values are derived from Figure 1, Table 1, and Figure 2 of Moller and Hough.[226]

(0.25 *M* sucrose containing 1 to 10 m*M* Tris per HCl). By varying the ionic composition of such media Moller and Hough[226] have shown that the enzyme appears to behave as an "ambiquitous" enzyme.[257] Figure 16 shows that the percentage of the Mg^{2+}-dependent phosphohydrolase activity in an adipocyte postmitochondrial supernatant that sediments at 106,000 *g* is substantially increased when KCl is incorporated into the homogenization buffer. This effect appears to be particularly sensitive to [KCl] in the range 0 to 20 m*M*. This phenomenon was interpreted as being an association of the soluble enzyme with the microsomal membranes although the experimental procedure does not discriminate this possibility from the alternative that salt may simply cause the soluble enzyme to undergo some form of aggregation. This membrane association or enzyme-aggregation phenomenon was reversible, was also promoted by low concentrations of Mg^{2+}, Ca^{2+} or spermine, and was diminished by increasing $[H^+]$ or by raising the temperature above 8 to 10°C. Aggregation or membrane association of the enzyme due to increased [KCl] was also observed when soluble fraction was mixed with mitochondria or plasma membranes suggesting either that phosphohydrolase will bind nonselectively to any cellular membrane or again that the phenomenon is simply an aggregation of the enzyme that does not require membranes at all. In this regard it is noteworthy that Moller and Hough[226] state that no phosphohydrolase activity was detectable in soluble fractions following incubation with Mg^{2+} or spermine and that this appeared to be due to precipitation and inactivation of the soluble enzyme. Jamdar and Osborne[230] have also reported that low concentrations of spermine added to adipose tissue postmitochondrial supernatants cause an increase in the microsomal phosphohydrolase activity and a decrease in the soluble fraction. These authors[230] have interpreted these findings more in terms of effects of spermine upon the enzyme activity (see Section III.D.8) than in terms of translocation between cell compartments. In the author's laboratory addition of spermine to 105,000-*g* supernatants from adipocytes substantially increases the proportion of the Mg^{2+}-dependent phosphohydrolase activity that is retained by a 0.3-μm pore-size filter.[267] This phenomenon is observed in the absence of microsomal, mitochondrial, or plasma membranes and appears to indicate promotion of enzyme aggregation by the polyamine.

Brindley[258] has suggested that controlled translocation of the cytosolic phosphohydrolase to the endoplasmic reticulum in liver represents a way in which an inactive reserve of the enzyme can be brought into action at the site of triacylglycerol synthesis. In accord with this idea Moller and Hough[226] have shown that microsomal fractions isolated from post-mitochondrial supernatants containing increased [KCl] have both an increased phosphohy-

Table 12
CHANGES IN SUBCELLULAR LOCALIZATION WITHIN RAT
ADIPOCYTES OF PHOSPHOHYDROLASE ACTIVITIES ON INCUBATION
WITH LIPOLYTIC AGENTS[20]

Incubation time (min)	Lipolytic agents	Fraction of enzyme in microsomes in treated cells/fraction of enzyme in microsomes in untreated cells × 100	
		Mg^{2+}-independent phosphohydrolase	Mg^{2+}-dependent phosphohydrolase
10	Adrenaline (5 μM)	98	187
10	Isoprenaline (5 μM)	108	242
10	Corticotropin (4.8 ng/mℓ)	120	134
20	cAMP (10 mM)	—	209
20	Theophylline (2.5 mM)	—	251
20	cAMP (10 mM) + theophylline (2.5 mM)	—	306
30	Dibutyryl cAMP (0.5 mM)	—	335

Note: Postmicrosomal supernatants were centrifuged at 160,000 *g* to obtain microsomal and soluble fractions. In cells incubated without lipolytic agents the average fraction of enzyme in the microsomes was 50% for Mg^{2+}-independent phosphohydrolase and 12% for Mg^{2+}-dependent phosphohydrolase.

drolase specific activity and, when incubated with [^{14}C]-glycerolphosphate, incorporate an increased proportion of the label into neutral lipids compared with that found in phosphatidate. However, again, this does not preclude the possibility that aggregation of the soluble phosphohydrolase might have occurred with a resulting enrichment of the activity in the microsomal fraction when this is isolated.

A cell-free system from rat liver translocation of Mg^{2+}-dependent phosphohydrolase to the endoplasmic reticulum has been demonstrated when long chain fatty acids or their CoA thioesters are added.[259-261] Using similar methodology with recombined adipocyte cytosol and microsomal membranes it has not yet been possible to demonstrate such an effect.[267] The possible involvement of fatty acids in control of the intracellular localization of the adipocyte phosphohydrolase is discussed in Sections VI.B and C.

B. Rapid Effects of Lipolytic Agents and Insulin on Phosphohydrolase Localization in Adipocytes In Vitro

Table 12 summarizes experimental findings of Moller et al.[20] who incubated rat adipocytes with various lipolytic agents for short times, prepared postmitochondrial supernatants, and then fractionated these by centrifugation at 160,000 *g* to yield microsomal and soluble fractions. Treatment of the cells had little effect on the total Mg^{2+}-dependent phosphohydrolase activity in the postmitochondrial supernatants but changed its distribution. In studies in this laboratory[21] it is also found that total Mg^{2+}-dependent phosphohydrolase in postmitochondrial supernatants obtained from noradrenaline-treated cells is not appreciably changed relative to the controls if the cells are not freeze-stopped (see Section IV.A.1) i.e., the rapid changes detected using the freeze-stop and rapid assay appear to relax rapidly. In the study of Moller et al.[20] extracts from untreated cells had only approximately 12% of the Mg^{2+}-dependent phosphohydrolase associated with the microsomes but his could be increased by treatment with lipolytic agents to as much as 40% in some instances. This "translocation" effect of catecholamine lipolytic agents appears to be mediated via the β-adrenoceptor. Table 12 shows that this change in intracellular localization was not seen with the Mg^{2+}-independent phosphohydrolase. An effect of noradrenaline on enzyme distribution has also been dem-

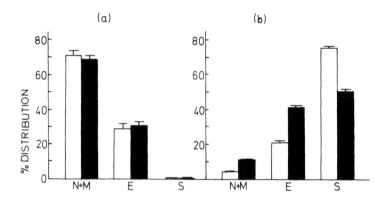

FIGURE 17.   Effect of noradrenaline on the subcellular distribution of phosphatidate phosphohydrolase activities in rat adipocytes. The values are derived
from Taylor and Saggerson[21] and are means ± SEM of 5 separate experiments.
Cells were incubated for 30 min in media containing 5 mM glucose and albumin
(40 mg/mℓ) with (closed histograms) or without (open histograms) noradrenaline
(0.5 μM) followed by gentle disruption of the cells and centrifugal fractionation.
N + M = nuclear + mitochondrial fraction; E = microsomal fraction; S =
105,000-g supernatant. Distribution of NADP$^+$-cytochrome c reductase was N
+ M, 27%; E, 64%; S, 9%; distribution of lactate dehydrogenase was N + M,
5%; E, 1%; S, 94%. The distribution of either of these markers was not affected
by noradrenaline treatment. (a) Mg^{2+}-independent phosphohydrolase; (b) Mg^{2+}-
dependent phosphohydrolase.

onstrated by Taylor and Saggerson[21] and Figure 17 again demonstrates that the effect is
confined to the Mg^{2+}-dependent activity. Furthermore, the increase in the particulate activity
is essentially confined to the microsomal fraction. The small increase in the proportion of
the activity that is in the nuclear and mitochondrial fraction can be attributed to cross
contamination by microsomes. Taylor and Saggerson[21] consistently found a higher proportion
of particulate Mg^{2+}-dependent phosphohydrolase than Moller et al.[20] This difference may
either be due to the use of higher ionic strength homogenization buffers (10 mM rather than
1 mM Tris-HCl in the 0.25 M sucrose) or may be due to differences in the procedures used
to distrupt the cells. Taylor and Saggerson[21] have also developed an alternative procedure
of fractionation whereby adipocytes are gently disrupted, the homogenates defatted by flo-
tation, and particulate and soluble phosphohydrolase separated by passage of the homogenate
through a 0.3-μm pore-size filter. Using this procedure, which avoids the hydrostatic pres-
sures of ultracentrifugation, approximately 40% of the Mg^{2+}-dependent phosphohydrolase
is found to be particulate in untreated cells and this is increased to approximately 70% after
incubation with noradrenaline.[21] This translocation is extremely rapid being fully established
within 10 min[21] and shows an identical noradrenaline dose dependence to that of activation
of lipolysis in that EC$_{50}$ values were 130 and 80 nM noradrenaline for phosphohydrolase
translocation and glycerol release respectively.[21] The close correlation between these two
events in each individual incubation of cells is indicated by Figure 18. Reference to Figures
4 and 5 (see Section II.D) shows that noradrenaline treatment has little effect on the kinetic
properties of the phosphohydrolase activity that remains in the soluble fraction. There is a
slight decrease in the apparent K$_m$ for phosphatidate (Figure 4) but no appreciable change
in the Mg^{2+} dependence of the soluble enzyme. In view of studies with the liver
phosphohydrolase[231,259-262] suggesting a role for fatty acids in promoting this translocation
it was considered possible that this effect of noradrenaline in adipocytes might be secondary
to stimulation of lipolysis. Adipocyte intracellular fatty acid concentration rises rapidly after
addition of noradrenaline and precedes the steady accumulation of fatty acids in the extra-

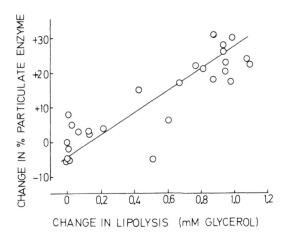

FIGURE 18. Relationship between activation of lipolysis and translocation of Mg^{2+}-dependent phosphohydrolase to the particulate fraction on incubation of rat adipocytes with noradrenaline. The values are derived from Taylor and Saggerson[21] and are each taken from individual incubations of cells. These were for 30 min in media containing 5 mM glucose, albumin (40 mg/mℓ), and varying concentrations of noradrenaline (0.01 to 10 μM). Separation of soluble and particulate phosphohydrolase was achieved by filtration of defatted homogenates through a 0.3-μm pore-size filter. The values are expressed relative to control incubations without noradrenaline in which glycerol release was negligible and the average proportion of the phosphohydrolase that was particulate was 43%. The regression line through the points has an r value of 0.865.

cellular medium.[263-265] In accord with this possibility, addition of palmitate to adipocyte incubations increased the proportion of the phosphohydrolase that was particulate and, furthermore, inclusion of albumin in homogenization buffers decreased this.[21] However, Figure 19 shows that albumin did not abolish the effect of noradrenaline. The percentage of the total by which the particulate phosphohydrolase was increased by noradrenaline was 24 to 25% whether or not albumin was present in homogenization buffers. This finding does not preclude a role for fatty acids in the translocation phenomenon but indicates that other factors also must mediate the effect of noradrenaline. One possibility is that the translocation may be dependent on the elevation of cAMP level and the activation of protein kinase that rapidly follows occupation of the β-adrenoceptor (see Section II.B). If this were so, the translocation response would be opposite to that seen in hepatocytes where cAMP[231] and glucagon[253] decrease the proportion of membrane-associated phosphohydrolase (Chapter 2, Section IV.E). Another possibility is that some of the noradrenaline-induced translocation (or perhaps aggregation) of the enzyme may be due to redistributions of ions. In this regard the sensitivity of the enzyme to ionic concentrations in cell-free systems is pertinent (see Section VI.A).

Insulin also affects the intracellular distribution of the adipocyte Mg^{2+}-dependent phosphohydrolase in that it opposes the effect of noradrenaline and by itself decreases the proportion of the enzyme that is particulate when cells are incubated without fatty acid (Figure 20). Again, this situation is opposite to that seen with hepatocytes where insulin increases the porportion of membrane-associated phosphohydrolase when fatty acid concentrations are low.[253] When adipocytes are incubated with glucose and palmitate, i.e., all

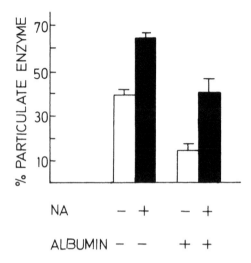

FIGURE 19.   Effect of albumin in extraction buffers on intracellular distribution of Mg^{2+}-dependent phosphohydrolase. The values are derived from Taylor and Saggerson[21] in an experiment in which rat adipocytes were incubated for 30 min with or without 0.5 μM noradrenaline (NA) followed by extraction of the cells in 0.25 nM sucrose-based media or the same also containing albumin (10 mg/mℓ) and separation of particulate and soluble enzyme by filtration (see Figure 18).

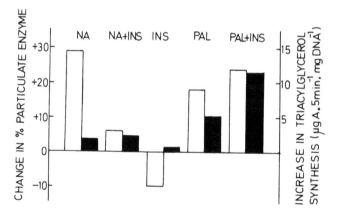

FIGURE 20.   Comparison of changes in intracellular distribution of Mg^{2+}-dependent phosphohydrolase activity with rates of triacylglycerol synthesis in rat adipocytes. The values are derived from Taylor and Saggerson[21] and show changes relative to the control of the particulate proportion of the phosphohydrolase (measured by a filtration procedure) and of the rate of [U14C]-glucose incorporation into glyceride glycerol during the final 5 min prior to sampling of the cells. Incubations were for 30 min and contained the following additions: 0.5 μM noradrenaline (NA), 3 nM insulin (INS), and 3 mM palmitate (PAL). Open histograms: change in percentage of the phosphohydrolase that is particulate (control = 45%); filled histograms: change in rate of triacylglycerol synthesis (control = 0.43 μg atoms/5 min/mg cell DNA).

the precursors are present to sustain a high rate of triacylglycerol synthesis (see Section II.A.2.a), insulin does not decrease the particulate proportion of the enzyme (Figure 20) and it may be seen that in the presence of insulin the degree of translocation of the enzyme brought about by palmitate is amplified (Figure 20).

What physiological significance can be attributed to these changes? Clearly, as shown by Figure 20, there is no general correlation between changes in the proportion of the enzyme that is particulate and changes in the rate of flux through the triacylglycerol synthesis pathway suggesting that, with regard to the overall regulation of the pathway, control strength must also be assigned to other components besides the phosphohydrolase. This is hardly surprising in view of the array of other regulatory processes described in Section II.A.2. However, it is striking that there is a high proportion of particulate enzyme when palmitate and insulin are present — a situation which should mimic that in vivo when lipoprotein lipase is active and the adipocyte is assimilating lipoprotein-derived fatty acids (see Section II.A.2.a.i). Finally, why should the particulate proportion of the enzyme be increased under lipolytic conditions when the obvious objective is to achieve release of fatty acids to the circulation? It must be remembered however, that the rapid translocation in response to noradrenaline or other lipolytic agents is accompanied by a time-dependent inactivation of the phosphohydrolase (see Section IV.A.1). It is not known if this inactivation is equally expressed in soluble and membrane-associated enzyme since at present it cannot be measured in the separated fractions. However, if it were, then at the onset of noradrenaline stimulation of the cell there would be a transient buffering or even a small increase in the microsomal phosphohydrolase activity but a considerable initial decline in the soluble activity. On the assumption that the membrane-associated enzyme is the metabolically active component this would facilitate some recycling back into triacylglycerol of mobilized fatty acids in the initial phase. Later as the total phosphohydrolase activity continued to fall, the microsomal phosphohydrolase activity would decline. In this later phase less of the mobilized fatty acid would be recycled and more would be released from the tissue. By this, the regulatory properties of the phosphohydrolase could contribute to achieving a more gradual build-up of fatty acid release from the tissue.

C. Changes in Phosphohydrolase Intracellular Localization In Vivo

In addition to the longer-term decreases in whole tissue Mg^{2+}-dependent phosphohydrolase activity seen in fasting and diabetes (see Sections V.A and B) the proportion of the enzyme that is particulate also changes in these states.[21] Of necessity these studies must be made in extracts from whole tissues which are removed from the animal and homogenized as rapidly as possible. The proportion of the enzyme that is particulate in these extracts is lower than that seen in extracts from adipocytes possibly due to the shear forces involved in homogenization of the whole tissue.[21] Although starvation and diabetes decreased the total tissue activity of the enzyme by 50% (see Table 11 and Figure 15) these states were accompanied by increases in the proportion of the remaining activity in the microsomal fraction (Figure 21).[21] Insulin reversed this translocation phenomenon in the diabetic state. The outcome again is that the microsomal phosphohydrolase activity is buffered. In the same study[21] the contralateral epididymal fat pads were homogenized in medium containing albumin. In these extracts the proportion of particulate enzyme was diminished relative to the albumin-free homogenates and it may be inferred that differing proportions of the microsomal activity may be in that particular subcellular fraction due to differing degrees of fatty acid-promoted translocation of the enzyme. At present this is the only study to show that the translocation phenomenon might occur in adipose tissue in vivo.

VII. CONCLUDING REMARKS

It is clear that the Mg^{2+}-dependent phosphohydrolase in white adipose tissue shows

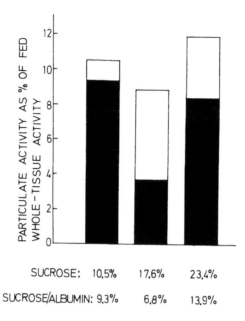

SUCROSE: 10.5%   17.6%   23.4%

SUCROSE/ALBUMIN: 9.3%   6.8%   13.9%

FIGURE 21. Particulate Mg^{2+}-dependent phospho-hydrolase activity in rat epididymal fat pads in starvation and diabetes. The measurements were made in the same experiment as Figure 15. The particulate proportion of the enzyme was that which sedimented at 105,000-g and the values for this are indicated below the abcissa for extracts prepared in 0.25 M sucrose-based homogenization media (open bars) or in the same also containing albumin (10 mg/mℓ (closed bars)). The values on the ordinate indicate these particulate activities calculated as a percentage of the control whole-tissue activity for the two extraction procedures. F = fed, S = 48-hr starved, D = streptozotocin-diabetic.

complex regulatory properties both short term and long term. In most instances these properties can be correlated with and integrated into the overall picture of the regulation of adipose tissue metabolism. It is probably simplistic to assign a dominant role to any regulatory enzyme in the overall control of any metabolic process. Rather, it is better to think in terms of the close-knit interplay between several regulatory enzymes. At the present time this must be considered to be the situation regarding the role of the phosphohydrolase partly because our knowledge is still incomplete and partly because it is clear that other enzymes in the triacylglycerol synthesis pathway are also regulated.

It is hoped that future studies will investigate the role of the phosphohydrolase in brown adipose tissue and that the enzyme will be investigated in human adipose tissue. An important advance that is awaited is the purification of the enzyme leading to the production of antibodies, determination of its composition, and structure and study of the genetic regulation of its activity.

ACKNOWLEDGMENTS

I am indebted to Dr. S. C. Jamdar for a sight of some of his work prior to publication and to Dr. C. H. K. Cheng and Dr. S. J. Taylor for being able to cite unpublished work from their doctorate theses. Work in my laboratory on the phosphohydrolase and triacylglycerol synthesis in general has been supported by the Medical Research Council (London) and the British Diabetic Association.

REFERENCES

1. **Slavin, B. G.,** The cytophysiology of mammalian adipose cells, *Int. Rev. Cytol.,* 33, 297, 1972.
2. **Greenwood, M. R. C. and Johnson, R. R.,** The adipose tissue, in *Histology, Cell and Tissue Biology,* Weiss, L., Ed., Elsevier, Amsterdam, 1983, 178.
3. **Salvin, B. G.,** The morphology of adipose tissue, in *New Perspectives in Adipose Tissue: Structure, Function and Development,* Cryer, A. and Van, R. L. R., Eds., Butterworths, London, 1985, 23.
4. **Nedergaard, J. and Lindberg, O.,** The brown fat cell, *Int. Rev. Cytol.,* 74, 187, 1982.
5. **Nicholls, D. and Locke, R.,** Cellular mechanisms of heat dissipation, in *Mammalian Thermogenesis,* Griardier, L. and Stock, M. J., Eds., Chapman and Hall, London, 1983, 8.
6. **Nicholls, D. G. and Locke, R. M.,** Thermogenic mechanisms in brown fat, *Physiol. Rev.,* 64, 1, 1984.
7. **Cannon, B. and Nedergaard, J.,** The biochemistry of an inefficient tissue: brown adipose tissue, in *Essays in Biochemistry,* Vol. 20, Campbell, P. N. and Marshall, R. D., Eds., Academic Press, Orlando, Fla., 1985, 110.
8. **Bates, E. J., Topping, D. L., Sooranna, S. R., Saggerson, E. D., and Mayes, P. A.,** Acute effects of insulin in glycerol phosphate acyltransferase activity, ketogenesis and serum free fatty acid concentration in perfused rat liver, *FEBS Lett.,* 84, 225, 1977.
9. **Bates, E. J. and Saggerson, E. D.,** A study of the glycerol phosphate acyltransferase and dihydroxyacetone phosphate acyltransferase activities in rat liver mitochondrial and microsomal fractions, *Biochem. J.,* 182, 751, 1979.
10. **Saggerson, E. D., Sooranna, S. R., and Cheng, C. H. K.,** Rapid hormonal control of enzymes of triacylglycerol synthesis in rat adipocytes, *INSERM Colloq.,* 87, 223, 1979.
11. **Saggerson, E. D., Carpenter, C. A., Cheng, C. H. K., and Sooranna, S. R.,** Subcellular distribution and some properties of N-ethyl-maleimide-sensitive and -insensitive forms of glycerol phosphate acyltransferase in rat adipocytes, *Biochem. J.,* 190, 183, 1980.
12. **Rider, M. H. and Saggerson, E. D.,** Regulation by noradrenaline of the mitochondrial and microsomal forms of glycerol phosphate acyltransferase in rat adipocytes, *Biochem. J.,* 214, 235, 1983.
13. **Dodds, P. F., Brindley, D. N., and Gurr, M. I.,** The effects of diet on the esterification of glycerol phosphate, dihydroxyacetone phosphate and 2-hexadecylglycerol by homogenates of rat adipose tissue, *Biochem. J.,* 160, 701, 1976.
14. **Sooranna, S. R. and Saggerson, E. D.,** Effects of starvation and adrenaline on glycerophosphate acyltransferase and dihydroxyacetone phosphate acyltransferase activities in rat adipocytes, *FEBS Lett.,* 99, 67, 1979.
15. **Schlossman, D. M. and Bell, R. M.,** Triacylglycerol synthesis in isolated fat cells. Evidence that the sn-glycerol-3-phosphate and dihydroxyacetone phosphate acyltransferase activities are dual catalytic functions of a single microsomal enzyme, *J. Biol. Chem.,* 251, 5738, 1976.
16. **Jones, C. L. and Hajra, A. K.,** The subcellular distribution of acyl CoA: dihydroxyacetone phosphate acyltransferase in guinea pig liver, *Biochem. Biophys. Res. Commun.,* 76, 1138, 1977.
17. **Declercq, P. E., Haagsman, H. P., Van Veldhoven, P., Debeer, L. J., Van Golde, L. M. G., and Mannaerts, G. P.,** Rat liver dihydroxyacetone-phosphate acyltransferases and their contribution to glycerolipid synthesis, *J. Biol. Chem.,* 259, 9064, 1984.
18. **Coleman, R. and Bell, R. M.,** Triacylglycerol synthesis in isolated fat cells. Studies on the microsomal diacylglycerol acyltransferase activity using ethanol-dispersed diacylglycerols, *J. Biol. Chem.,* 251, 4537, 1976.
19. **Jamdar, S. C. and Fallon, H. J.,** Glycerolipid synthesis in rat adipose tissue. II. Properties and distribution of phosphatidate phosphatase, *J. Lipid Res.,* 14, 517, 1973.
20. **Moller, F., Wong, K. H., and Green, P.,** Control of fat cell phosphatidate phosphohydrolase by lipolytic agents, *Can. J. Biochem.,* 59, 9, 1981.
21. **Taylor, S. J. and Saggerson, E. D.,** Adipose tissue Mg^{2+}-dependent phosphatidate phosphohydrolase. Control of activity and subcellular distribution *in vitro* and *in vivo*, *Biochem. J.,* in press.
22. **Yamashita, S., Hosaka, K., Taketo, M., and Numa, S.,** Distribution of glycerolipid-synthesizing enzymes in the subfractions of rat liver microsomes, *FEBS Lett.,* 29, 235, 1973.
23. **Lamb, R. G. and Fallon, H. J.,** Glycerolipid formation from sn-glycerol-3-phosphate by rat liver cell fractions. The role of phosphatidate phosphohydrolase, *Biochim. Biophys. Acta,* 348, 166, 1974.
24. **Dang, A. -Q., Faas, F. H., and Carter, W. J.,** Effects of streptozotocin-induced diabetes on phosphoglyceride metabolism of the rat liver, *Lipids,* 19, 738, 1984.
25. **Dang, A. -Q., Faas, F. H., and Carter, W. J.,** Influence of hypo- and hyperthroidism on rat liver glycerolipid metabolism, *Lipids,* 20, 897, 1985.
26. **Young, D. L. and Lynen, F.,** Enzymatic regulation of 3-sn-phosphatidylcholine and triacylglycerol synthesis in states of altered lipid metabolism, *J. Biol. Chem.,* 244, 377, 1969.

27. **Saggerson, E. D. and Tomassi, G.,** The regulation of glyceride synthesis from pyruvate in isolated fat cells. The effects of palmitate and alteration of dietary status, *Eur. J. Biochem.,* 23, 109, 1971.

28. **Saggerson, E. D.,** The regulation of glyceride synthesis in isolated white-fat cells. The effects of palmitate and lipolytic agents, *Biochem. J.,* 128, 1057, 1972.

29. **Saggerson, E. D.,** Hormonal regulation of biosynthetic activities in white adipose tissue, in *New Perspectives in Adipose Tissue: Structure, Function and Development,* Cryer, A. and Van, R. L. R., Eds., Butterworths, London, 1985, 87.

30. **Hems, D. A., Rath, E. A., and Verrinder, T. R.,** Fatty acid synthesis in liver and adipose tissue of normal and genetically obese (ob/ob) mice during the 24-hour cycle, *Biochem. J.,* 150, 167, 1975.

31. **Katz, J. and Wals, P. A.,** Lipogenesis from lactate in rat adipose tissue, *Biochim. Biophys. Acta,* 348, 344, 1974.

32. **Rath, E. A., Beloff-Chain, A., and Hems, D. A.,** Contribution of lactate carbon to fatty acid synthesis in adipose tissue of normal and genetically obese (ob/ob) mice, *Biochem. Soc. Trans.,* 3, 513, 1975.

33. **Ballard, F. J., Hanson, R. W., and Leveille, G. A.,** Phosphoenolpyruvate carboxykinase and the synthesis of glyceride-glycerol from pyruvate in adipose, *J. Biol. Chem.,* 242, 2746, 1967.

34. **Reshef, L., Hanson, R. W., and Ballard, F. J.,** Glyceride-glycerol synthesis from pyruvate. Adaptive changes in phosphoenolpyruvate carboxykinase and pyruvate carboxylase in adipose tissue and liver, *J. Biol. Chem.,* 244, 1994, 1969.

35. **Gorin, E., Tal-Or, Z., and Shafrir, E.,** Glyceroneogenesis in adipose tissue of fasted, diabetic and tiramcinolone treated rats, *Eur. J. Biochem.,* 8, 370, 1969.

36. **Chakrabarty, K. and Leveille, G. A.,** Conversion of pyruvate to glyceride glycerol in adipose tissue of obese and nonobese mice, *Arch. Biochem. Biophys.,* 125, 259, 1968.

37. **Saggerson, E. D.,** The regulation of glyceride synthesis in isolated white-fat cells. The effects of acetate, pyruvate, lactate, palmitate, electron-acceptors, uncoupling agents and oligomycin, *Biochem. J.,* 128, 1069, 1972.

38. **Sooranna, S. R. and Saggerson, E. D.,** Studies on the role of insulin in the regulation of glyceride synthesis in rat epididymal adipose tissue, *Biochem. J.,* 150, 441, 1975.

39. **Saggerson, E. D. and Greenbaum, A. L.,** The regulation of triglyceride synthesis and fatty acid synthesis in rat epididymal adipose tissue. Effects of insulin, adrenaline and some metabolities in vitro, *Biochem. J.,* 119, 193, 1970.

40. **Harper, R. D. and Saggerson, E. D.,** Factors affecting fatty acid oxidation in fat cells isolated from rat white adipose tissue, *J. Lipid Res.,* 17, 516, 1976.

41. **Smith, S. J. and Saggerson, E. D.,** Regulation of pyruvate dehydrogenase activity in rat epididymal fat pads and isolated adipocytes by adrenaline, *Biochem. J.,* 174, 119, 1979.

42. **McGarry, J. D., Meier, J. M., and Foster, D. W.,** The effects of starvation and refeeding on carbohydrate and lipid metabolism *in vivo* and in the perfused rat liver. The relationship between fatty acid oxidation and esterification in the regulation of ketogenesis, *J. Biol. Chem.,* 248, 270, 1973.

43. **Saggerson, E. D.,** Regulation of lipid metabolism in adipose tissue and liver cells, in *Biochemistry of Cellular Regulation,* Vol. 2, Ashwell, M., Ed., CRC Press, Boca Raton, Fla., 1980, 207.

44. **Ide, T. and Ontko, J. A.,** Increased secretion of very low density lipoprotein triglyceride following inhibition of long chain fatty acid oxidation in isolated rat liver, *J. Biol. Chem.,* 256, 10247, 1981.

45. **Cryer, A., Davies, P., Williams, E. R., and Robinson, D. S.,** The clearing factor lipase activity of isolated fat cells, *Biochem. J.,* 146, 481, 1975.

46. **Cryer, A., McDonald, A., Williams, E. R., and Robinson, D. S.,** Colchicine inhibition of the heparin-stimulated release of clearing factor lipase from isolated fat cells, *Biochem. J.,* 152, 717, 1975.

47. **Cryer, A.,** Tissue lipoprotein lipase activity and its action in lipoprotein metabolism, *Int. J. Biochem.,* 13, 525, 1981.

48. **Robinson, D. S.,** The function of the plasma triglycerides in fatty acid transport, in *Comprehensive Biochemistry,* Florkin, M. and Stotz, E. H., Eds., Elsevier, Amsterdam, 1970, 51.

49. **Garfinkel, A. S. and Schotz, M. C.,** Sequential induction of two species of lipoprotein lipase, *Biochim. Biophys. Acta,* 306, 128, 1973.

50. **Pav, J. and Wenkeova, J.,** Significance of adipose tissue lipoprotein lipase, *Nature (London),* 185, 926, 1960.

51. **Kessler, J. I.,** Effects of diabetes and insulin on the activity of myocardial and adipose tissue lipoprotein lipase in rats, *J. Clin. Invest.,* 42, 362, 1963.

52. **Brown, D. F., Dandiss, K., and Durrant, J.,** Triglyceride metabolism in the alloxan-diabetic rat, *Diabetes,* 16, 90, 1967.

53. **Otway, S. and Robinson, D. S.,** The significance of changes in tissue clearing-factor lipase activity in relation to the lipaemia of pregnancy, *Biochem. J.,* 106, 677, 1968.

54. **Hamosh, M., Clary, T. R., Chernick, S. S., and Scow, R. O.,** Lipoprotein lipase activity of adipose and mammary tissue and plasma triglyceride in pregnant and lactating rats, *Biochim. Biophys. Acta,* 210, 473, 1970.

55. **Cryer, A., Riley, S. E., Williams, E. R., and Robinson, D. S.,** Effects of fructose, sucrose and glucose feeding on plasma insulin concentrations and on adipose-tissue clearing-factor lipase activity in the rat, *Biochem. J.,* 140, 561, 1974.
56. **Reichl, D.,** Lipoprotein lipase activity in the adipose tissue of rats adapted to controlled feeding schedules, *Biochem. J.,* 128, 79, 1972.
57. **Salaman, M. R. and Robinson, D. S.,** Clearing factor lipase in adipose tissue, a medium in which the enzyme activity of tissue from starved rats increases *in vitro, Biochem. J.,* 99, 640, 1966.
58. **Wing, D. R., Salaman, M. R., and Robinson, D. S.,** Clearing factor lipase in adipose tissue. Factors influencing the increase in enzyme activity produced on incubation of tissue from starved rats *in vitro, Biochem. J.,* 99, 648, 1966.
59. **Spooner, P. M., Chernick, S. S., Garrison, M. M., and Scow, R. O.,** Insulin regulation of lipoprotein lipase activity and release in 3T3-L1-adipocytes. Separation and dependence of hormonal effects on hexose metabolism and synthesis of RNA and protein, *J. Biol. Chem.,* 254, 10021, 1979.
60. **Wing, D. R. and Robinson, D. S.,** Clearing factor lipase in adipose tissue. Studies with puromycin and actinomycin, *Biochem. J.,* 106, 667, 1968.
61. **Nestel, P. J. and Austin, W.,** Relationship between adipose lipoprotein lipase activity and compounds which affect intracellular lipolysis, *Life Sci.,* 8, 157, 1969.
62. **Davies, P., Cryer, A., and Robinson, D. S.,** Hormonal control of adipose tissue clearing factor lipase activity, *FEBS Lett.,* 45, 271, 1974.
63. **Ashby, P., Bennett, D. P., Spencer, I. M., and Robinson, D. S.,** Post-translational regulation of lipoprotein lipase activity in adipose tissue, *Biochem. J.,* 176, 865, 1978.
64. **Ashby, P. and Robinson, D. S.,** Effect of insulin, glucocorticoids and adrenaline on the activity of rat adipose tissue lipoprotein lipase, *Biochem. J.,* 188, 185, 1980.
65. **Cryer, A.,** Lipoproteins and adipose tissue, in *New Perspectives in Adipose Tissue: Structure, Function and Development,* Cryer, A. and Van, R. L. R., Eds., Butterworths, London, 1985, 183.
66. **Flatt, J. P. and Ball, E. G.,** Studies on the metabolism of adipose tissue. XV. An evaluation of the major pathways of glucose catabolism as influenced by insulin and epinephrine, *J. Biol. Chem.,* 239, 675, 1964.
67. **Katz, J., Landau, B. R., and Bartsch, G. E.,** The pentose cycle, triose phosphate isomerization and lipogenesis in rat adipose tissue, *J. Biol. Chem.,* 244, 727, 1966.
68. **Saggerson, E. D. and Greenbaum, A. L.,** The regulation of triglyceride synthesis and fatty acid synthesis in rat epididymal adipose tissue. Effects of altered dietary and hormonal conditions, *Biochem. J.,* 119, 221, 1970.
69. **Simpson, I. A. and Cushman, S. W.,** Regulation of glucose transporter and hormone receptor cycling by insulin in the rat adipose cell, *Curr. Top. Membranes Transport,* 24, 459, 1985.
70. **Denton, R. M. and Hughes, W. A.,** Pyruvate dehydrogenase and the hormonal regulation of fat synthesis in mammalian tissues, *Int. J. Biochem.,* 9, 545, 1978.
71. **Denton, R. M., Brownsey, R. W., and Belsham, G. J.,** A partial view of the mechanism of insulin action, *Diabetologia,* 21, 347, 1981.
72. **Newsholme, E. A., Arch, J. R. S., Brooks, B., and Surholt, B.,** The role of substrate cycles in metabolic regulation, *Biochem. Soc. Trans.,* 11, 52, 1983.
73. **Abumrad, N. A., Perkins, R. C., Park, J. H., and Park, C. R.,** Mechanism of long chain fatty acid permeation in the isolated adipocyte, *J. Biol. Chem.,* 256, 9183, 1981.
74. **Abumrad, N. A., Perry, P. R., and Whitesell, R. R.,** Stimulation by epinephrine of the membrane transport of long chain fatty acid in the adipocyte, *J. Biol. Chem.,* 260, 9969, 1985.
75. **Crofford, O. B. and Renold, A. E.,** Glucose uptake by incubated rat epididymal adipose tissue. Rate-limiting steps and site of insulin action, *J. Biol. Chem.,* 240, 14, 1965.
76. **Vinten, J., Gliemann, J., and Østerlind, K.,** Exchange of 3-O-methyl-glucose in isolated fat cells. Concentration-dependence and effect of insulin, *J. Biol. Chem.,* 251, 794, 1976.
77. **Halperin, M. L., Mak, M. L., and Taylor, W. M.,** Control of glucose transport in adipose tissue of the rat: role of insulin, ATP and intracellular metabolites, *Can. J. Biochem.,* 56, 708, 1978.
78. **Livingston, J. N., Amatruda, J. M., and Lockwood, D. H.,** Studies of glucose transport system of fat cells: effects of insulin and insulin mimickers, *Am. J. Physiol.,* 234, E484, 1978.
79. **Olefsky, J. M.,** Mechanisms of the ability of insulin to activate the glucose-transport system in rat adipocytes, *Biochem. J.,* 172, 137, 1978.
80. **Whitesell, R. R. and Gliemann, J.,** Kinetic parameters of transport of 3-*O*-methylglucose and glucose in adipocytes, *J. Biol. Chem.,* 254, 2576, 1979.
81. **Czech, M. P.,** Insulin action and the regulation of hexose transport, *Diabetes,* 20, 399, 1980.
82. **Karnieli, E., Zarnowski, M. J., Hissin, P. J., Simpson, I. A., Salans, L. B., and Cushman, S. W.,** Insulin-stimulated translocation of glucose transport systems in the isolated rat adipose cell, *J. Biol. Chem.,* 256, 4772, 1981.
83. **Kashiwagi, A., Huecksteadt, T. P., and Foley, J. E.,** The regulation of glucose transport of cAMP stimulators via three different mechanisms in rat and human adipocytes, *J. Biol. Chem.,* 258, 13685, 1983.

84. **Smith, U., Kuroda, M., and Simpson, I. A.,** Counter-regulation of insulin-stimulated glucose transport by catecholamines in the isolated rat adipose cell, *J. Biol. Chem.,* 259, 8758, 1984.

85. **Whitesell, R. R. and Abumrad, N. A.,** Increased affinity predominates in insulin stimulation of glucose transport in the adipocyte, *J. Biol. Chem.,* 260, 2894, 1985.

86. **Denton, R. M., Yorke, R. E., and Randle, P. J.,** Measurement of concentrations of metabolites in adipose tissue and effects of insulin, alloxan-diabetes and adrenaline, *Biochem. J.,* 100, 407, 1966.

87. **Denton, R. M. and Halperin, M. L.,** The control of fatty acid and triglyceride synthesis in rat epididymal adipose tissue, *Biochem. J.,* 110, 27, 1968.

88. **Halperin, M. L. and Denton, R. M.,** Regulation of glycolysis and L-glycerol 3-phosphate concentration in rat epididymal adipose tissue *in vitro*. Role of phosphofructokinase, *Biochem. J.,* 113, 207, 1969.

89. **Cushman, S. W. and Wardzala, L. J.,** Potential mechanism of insulin action on glucose transport in the isolated rat adipose cell-apparent translocation of intracellular transport systems to the plasma membrane, *J. Biol. Chem.,* 255, 4758, 1980.

90. **Suzuki, K. and Kono, T.,** Evidence that insulin causes translocation of glucose transport activity to the plasma membrane from an intracellular storage site, *Proc. Natl. Acad. Sci. U.S.A.,* 77, 2542, 1980.

91. **Kono, T., Suzuki, K., Dansey, L. E., Robinson, F. W., and Blevins, T. L.,** Energy-dependent and protein synthesis-independent recycling of the insulin-sensitive glucose transport mechanism in fat cells, *J. Biol. Chem.,* 256, 6400, 1981.

92. **Carter-Su, C. and Czech, M. P.,** Reconstitution of D-glucose transport activity from cytoplasmic membranes. Evidence against recruitment of cytoplasmic membrane transporters into the plasma membrane as the sole action of insulin, *J. Biol. Chem.,* 255, 10382, 1980.

93. **Czech, M. P.,** Molecular basis of insulin action, *Ann. Rev. Biochem.,* 46, 359, 1977.

94. **Hyslop, P. A., Kuhn, C. E., and Sauerheber, R. D.,** Insulin stimulation of glucose transport in isolated rat adipocytes. Functional evidence for insulin activation of intrinsic transporter activity within the plasma membrane, *Biochem. J.,* 232, 245, 1985.

95. **Joost, H. -G. and Steinfelder, H. J.,** Insulin-like stimulation of glucose transport in isolated adipocytes by fatty acids, *Biochem. Biophys. Res. Commun.,* 128, 1358, 1985.

96. **Taylor, W. M. and Halperin, M. L.,** Stimulation of glucose transport in rat adipocytes by insulin, adenosine, nicotinic acid and hydrogen peroxide, *Biochem. J.,* 178, 381, 1979.

97. **Joost, H. G. and Steinfelder, H. J.,** Modulation of insulin sensitivity by adenosine. Effects on glucose transport, lipid synthesis and insulin receptors of the adipocyte, *Mol. Pharmacol.,* 22, 614, 1982.

98. **Denton, R. M. and Randle, P. J.,** Citrate and the regulation of adipose-tissue phosphofructokinase, *Biochem. J.,* 100, 420, 1966.

99. **Sooranna, S. R. and Saggerson, E. D.,** Rapid modulation of adipocyte phosphofructokinase activity by noradrenaline and insulin, *Biochem. J.,* 202, 753, 1982.

100. **Lederer, B. and Hess, H-G.,** On the mechanism by which noradrenaline increases the activity of phosphofructokinase in isolated rat adipocytes, *Biochem. J.,* 217, 709, 1984.

101. **Rider, M. H. and Hue, L.,** Regulation of fructose 2,6-biphosphate concentration in white adipose tissue, *Biochem. J.,* 225, 421, 1985.

102. **Sale, E. M. and Denton, R. M.,** β-Adrenergic agents increase the phosphorylation of phosphofructokinase in isolated rat epididymal white adipose tissue, *Biochem. J.,* 232, 905, 1985.

103. **Sobrino, F. and Gualberto, A.,** Hormonal regulation of fructose 2,6-biphosphate levels in epididymal adipose tissue of rat, *FEBS Lett.,* 182, 327, 1985.

104. **May, J. M.,** Triacylglycerol turnover in large and small rat adipocytes: effects of lipolytic stimulation, glucose and insulin, *J. Lipid Res.,* 23, 428, 1982.

105. **Taylor, W. M., Mak, M. L., and Halperin, M. L.,** Effect of 3':5'-cyclic AMP on glucose transport in rat adipocytes, *Proc. Natl. Acad. Sci. U.S.A.,* 73, 4359, 1976.

106. **Green, A.,** Catecholamines inhibit insulin-stimulated glucose transport in adipocytes in the presence of adenosine deaminase, *FEBS Lett.,* 152, 261, 1983.

107. **Green, A.,** Glucagon inhibition of insulin-stimulated 2-deoxyglucose uptake by rat adipocytes in the presence of adenosine deaminase, *Biochem. J.,* 212, 189, 1983.

108. **Schimmel, R. J. and Goodman, H. M.,** Effects of dibutyryl cyclic adenosine 3',5'-monophosphate on glucose transport and metabolism in rat adipose tissue, *Biochim. Biophys. Acta,* 239, 9, 1971.

109. **Ludvigsen, C., Jarett, L., and McDonald, J. M.,** The characterization of catecholamine-stimulation of glucose transport by rat adipocytes and isolated plasma membranes, *Endocrinology,* 106, 786, 1980.

110. **Leboeuf, B., Flinn, R. B., and Cahill, G. F.,** Effect of epinephrine on glucose uptake and glycerol release by adipose tissue in vitro, *Proc. Soc. Exp. Biol. Med.,* 102, 527, 1959.

111. **Lynn, W. S., MacLeod, R. M., and Brown, R. H.,** Effects of epinephrine, insulin and corticotropin on the metabolism of rat adipose tissue, *J. Biol. Chem.,* 235, 1904, 1960.

112. **Blecher, M.,** Evidence for the involvement of cyclic 3',5'-adenosine monophosphate in glucose utilization by isolated rat epididymal adipose tissue cells, *Biochem. Biophys. Res. Commun.,* 27, 560, 1967.

113. **Blecher, M., Merlino, N. S., Ro'ane, J. T., and Flynn, P. D.,** Independence of the effects of epinephrine, glucagon and adrenocorticotropin on glucose utilization from those on lipolysis in isolated rat adipose cells, *J. Biol. Chem.,* 244, 3423, 1969.

114. **Luzio, J. P., Jones, R. C., Siddle, K., and Hales, C. N.,** Dissociation of the effect of adrenaline on glucose uptake from rat on adenosine cyclic 3',5'-monophosphate levels and on lipolysis in rat isolated fat cells, *Biochim. Biophys. Acta,* 362, 29, 1974.

115. **Sale, E. M. and Denton, R. M.,** Adipose-tissue phosphofructokinase. Rapid purification and regulation by phosphorylation in vitro, *Biochem. J.,* 232, 897, 1985.

116. **Yorke, R. E.,** The influence of dexamethasone on adipose tissue metabolism *in vitro, J. Endocrinol.,* 39, 329, 1967.

117. **Livingston, J. N. and Lockwood, D. H.,** Effects of glucocorticoids on the glucose transport system of isolated fat cells, *J. Biol. Chem.,* 250, 8353, 1975.

118. **Olefsky, J. M.,** Effects of fasting on insulin binding, glucose transport, and glucose oxidation in isolated rat adipocytes: relationships between insulin receptors and insulin action, *J. Clin. Invest.,* 58, 1450, 1976.

119. **Karnieli, E., Hissin, P. J., Simpson, I. A., Salans, L. B., and Cushman, S. W.,** A possible mechanism of insulin resistance in the rat adipose cell in streptozotocin-induced diabetes mellitus, *J. Clin. Invest.,* 68, 811, 1981.

120. **Hissin, P., Foley, J. E., Wardzala, L. J., Karnieli, E., Simpson, I. A., Salans, L. B., and Cushman, S. W.,** Mechanism of insulin resistant glucose transport activity in the enlarged adipose cell of the aged, obese rat, *J. Clin. Invest.,* 70, 780, 1982.

121. **Hissin, P. J., Karnieli, E., Simpson, I. A., Salans, L. B., and Cushman, S. W.,** A possible mechanism of insulin resistance in the rat adipose cell with high fat/low carbohydrate feeding. Depletion of intracellular glucose transport systems, *Diabetes,* 31, 589, 1982.

122. **Shafrir, E. and Kerpel, S.,** Fatty acid esterification and release as related to the carbohydrate metabolism of adipose tissue: effect of epinephrine, cortisol and adrenalectomy, *Arch. Biochem. Biophys.,* 105, 237, 1964.

123. **Grahn, M. F. and Davies, J. I.,** Lipolytic agents as regulators of fatty acid esterification in rat adipose tissue, *Biochem. Soc. Trans.,* 8, 362, 1980.

124. **Sooranna, S. R. and Saggerson, E. D.,** Interactions of insulin and adrenaline with glycerol phosphate acylation processes in fat-cells from rat, *FEBS Lett.,* 64, 36, 1976.

125. **Sooranna, S. R. and Saggerson, E. D.,** Studies of the effects of adrenaline on glycerol phosphate acyltransferase activity in rat adipocytes, *FEBS Lett.,* 90, 141, 1978.

126. **Sooranna, S. R. and Saggerson, E. D.,** A stable decrease in long chain fatty acyl CoA synthetase activity after treatment of rat adipocytes with adrenaline, *FEBS Lett.,* 92, 241, 1978.

127. **Hall, M. and Saggerson, E. D.,** Reversible inactivation by noradrenaline of long-chain fatty acyl-CoA synthetase in rat adipocytes, *Biochem. J.,* 226, 275, 1985.

128. **Cheng, C. H. K. and Saggerson, E. D.,** Rapid effects of noradrenaline on Mg^{2+}-dependent phosphatidate phosphohydrolase activity in rat adipocytes, *FEBS Lett.,* 87, 65, 1978.

129. **Cheng, C. H. K. and Saggerson, E. D.,** Rapid antagonistic actions of noradrenaline and insulin on rat adipocyte phosphatidate phosphohydrolase activity, *FEBS Lett.,* 93, 120, 1978.

130. **Cheng, C. H. K. and Saggerson, E. D.,** The inactivation of rat adipocyte Mg^{2+}-dependent phosphatidate phosphohydrolase by noradrenaline, *Biochem. J.,* 190, 659, 1980.

131. **Cheng, C. H. K., Sooranna, S. R., and Saggerson, E. D.,** Effects of noradrenaline and $N^6,O^{2'}$-dibutyryl 3',5'-cyclic AMP on adipocyte glycerolipid-synthesizing enzymes, *Int. J. Biochem.,* 12, 667, 1980.

132. **Sooranna, S. R. and Saggerson, E. D.,** A decrease in diacylglycerol acyltransferase after treatment of rat adipocytes with adrenaline, *FEBS Lett.,* 95, 85, 1978.

133. **Jason, C. J., Polokoff, M. A., and Bell, R. M.,** Triacylglycerol synthesis in isolated fat cells. An effect of insulin on microsomal fatty acid coenzyme A ligase activity, *J. Biol. Chem.,* 251, 1488, 1976.

134. **Evans, G. L. and Denton, R. M.,** Regulation of fatty acid synthesis and esterification in rat epididymal adipose tissue: effects of insulin, palmitate and 2-bromopalmitate, *Biochem. Soc. Trans.,* 5, 1288, 1977.

135. **Nimmo, H. G. and Houston, B.,** Rat adipose tissue glycerol phosphate acyltransferase can be inactivated by cyclic AMP-dependent protein kinase, *Biochem. J.,* 176, 607, 1978.

136. **Nimmo, G. A. and Nimmo, H. G.,** Studies of rat adipose-tissue microsomal glycerol phosphate acyltransferase, *Biochem. J.,* 224, 101, 1984.

137. **Angel, A. and Roncari, D. A. K.,** The control of fatty acid esterification in a subcellular preparation of rat adipose tissue, *Biochim. Biophys. Acta,* 137, 464, 1967.

138. **Aas, M. and Daae, L. N. W.,** Fatty acid activation and acyl-transfer in organs from rats in different nutritional states, *Biochim. Biophys. Acta,* 239, 208, 1971.

139. **Jamdar, S. C. and Osborne, L. J.,** Glycerolipid biosynthesis in rat adipose tissue. Changes during a starvation and refeeding cycle, *Biochim. Biophys. Acta,* 713, 647, 1982.

140. **Himms-Hagen, J.,** Lipid metabolism in warm acclimated and cold acclimated rats exposed to cold, *Can. J. Physiol. Pharmacol.,* 43, 379, 1965.

141. **Fain, J. N., Reed, N., and Saperstein, R.,** The isolation and metabolism of brown fat cells, *J. Biol. Chem.,* 242, 1887, 1967.

142. **Angel, A.,** Brown adipose cells: spontaneous mobilization of endogenously synthesized lipid, *Science,* 163, 288, 1969.

143. **Knight, B. L. and Myant, N. B.,** A comparison between the effects of cold exposure *in vivo* and of noradrenaline *in vitro* on the metabolism of the brown fat of new-born rabbits, *Biochem. J.,* 119, 103, 1970.

144. **Knight, B. L. and Myant, N. B.,** The effect of noradrenaline on glyceride synthesis and oxidative metabolism *in vitro* in the brown fat of newborn rabbits, *Biochem. J.,* 125, 1, 1971.

145. **Paetzke-Brunner, I., Loffler, G., and Wieland, O. H.,** Activation of pyruvate dehydrogenase by insulin in isolated brown fat cells, *Horm. Metab. Res.,* 11, 285, 1979.

146. **Czech, M. P., Lawrence, J. C., and Lynn, W. S.,** Hexose transport in brown fat cells, *J. Biol. Chem.,* 249, 5421, 1974.

147. **Treble, D. H. and Ball, E. G.,** The occurrence of glycerokinase in rat brown adipose tissue, *Fed. Proc. Fed. Am. Soc. Exp. Biol.,* 22, 357, 1963.

148. **Prusiner, S. B., Cannon, B., Ching, T. M., and Lindberg, O.,** Oxidative metabolism in cells isolated from brown adipose tissue. Catecholamine regulated respiratory control, *Eur. J. Biochem.,* 7, 51, 1968.

149. **Schultz, F. M. and Johnston, J. M.,** The synthesis of higher glycerides via the monoglyceride pathway in hamster adipose tissue, *J. Lipid Res.,* 12, 132, 1971.

150. **Schenk, H., Heim, T., Mende, T., Varga, F., and Goetze, E.,** Studies on plasma free-fatty-acid metabolism and triglyceride synthesis of brown adipose tissue *in vivo* during cold-induced thermogenesis of the newborn rabbit, *Eur. J. Biochem.,* 58, 15, 1975.

151. **Pedersen, J. I., Slinde, E., Grynne, B., and Aas, M.,** The intracellular localization of long-chain acyl-CoA synthetase in brown adipose tissue, *Biochim. Biophys. Acta,* 398, 191, 1975.

152. **Normann, P. T. and Flatmark, T.,** Acyl-CoA synthetase activity of brown adipose tissue mitochondria, *Biochim. Biophys. Acta,* 619, 1, 1980.

153. **Lindberg, O., De Presse, J., Rylander, E., and Afzelius, B. A.,** Studies of the mitochondrial energy-transfer system of brown adipose tissue, *J. Cell Biol.,* 34, 293, 1967.

154. **Meisner, H. and Carter, J. R.,** Regulation of lipolysis in adipose tissue, *Horizons Biochem. Biophys.,* 4, 91, 1977.

155. **Hales, C. N., Luzio, J. P., and Siddle, K.,** Hormonal control of adipose-tissue lipolysis, *Biochem. Soc. Symp.,* 43, 97, 1978.

156. **Burns, T. W., Terry, B. E., Langley, P. E., and Robison, G. A.,** Role of cyclic AMP in human adipose tissue lipolysis, *Adv. Cyclic Nucleotide Res.,* 12, 329, 1980.

157. **Fain, J. N.,** Hormonal regulation of lipid mobilization from adipose tissue, in *Biochemical Action of Hormones,* Vol. 7, Litwack, G., Ed., Academic Press, New York, 1980, 119.

158. **Davis, J. T. and Souness, J. E.,** The mechanisms of hormone and drug actions on fatty acid release from adipose tissue, *Rev. Pure Appl. Pharmacol. Sci.,* 2, 1, 1981.

159. **Fain, J. N.,** Regulation of lipid metabolism by cyclic nucleotides, in *Handbook of Experimental Pharmacology,* Vol. 58/II, Nathanson, J. A. and Kebabian, J. W., Eds., Springer-Verlag, Berlin, 1982, 89.

160. **Belfrage, P.,** Hormonal control of lipid degradation, in *New Perspectives in Adipose Tissue: Structure, Function and Development,* Cryer, A. and Van, R. L. R., Eds., Butterworths, London, 1985, 121.

161. **Huttunen, J. K., Steinberg, D., and Mayer, S. E.,** ATP-dependent and cyclic AMP-dependent activation of rat adipose tissue lipase by protein kinase from rabbit skeletal muscle, *Proc. Natl. Acad. Sci. U.S.A.,* 67, 290, 1970.

162. **Huttunen, J. K., Steinberg, D., and Mayer, S. E.,** Protein kinase activation and phosphorylation of a purified hormone-sensitive lipase, *Biochem. Biophys. Res. Commun.,* 41, 1350, 1970.

163. **Huttunen, J. K., Ellingboe, J., Pittman, R. C., and Steinberg, D.,** Partial purification and characterization of hormone-sensitive lipase from rat adipose tissue, *Biochim. Biophys. Acta,* 218, 333, 1970.

164. **Tsai, S-C., Belfrage, P., and Vaughan, M.,** Activation of hormone-sensitive lipase in extracts of adipose tissue, *J. Lipid Res.,* 11, 466, 1970.

165. **Corbin, J. D., Reimann, E. M., Walsh, D. A., and Krebs, E. G.,** Activation of adipose tissue lipase by skeletal muscle cyclic adenosine 3′,5′-monophosphate-stimulated protein kinase, *J. Biol. Chem.,* 245, 4849, 1970.

166. **Khoo, J. C. and Steinberg, D.,** Reversible protein kinase activation of hormone-sensitive lipase from chicken adipose tissue, *J. Lipid Res.,* 15, 602, 1974.

167. **Khoo, J. C., Steinberg, D., Huang, J. J., and Vagelos, P. R.,** Triglyceride, diglyceride, monoglyceride and cholesterol ester hydrolase in chicken adipose tissue activated by adenosine 3′:5′-monophosphate-dependent protein kinase, *J. Biol. Chem.,* 251, 2882, 1976.

168. **Khoo, J. C., Steinberg, D., and Lee, E. Y. C.,** Activation of chicken adipose tissue diglyceride lipase by cyclic AMP-dependent protein kinase and its deactivation by purified protein phosphatase, *Biochem. Biophys. Res. Commun.,* 80, 418, 1978.

169. **Steinberg, D.,** Interconvertible enzymes in adipose tissue regulated by cyclic AMP-dependent protein kinase, *Adv. Cyclic Nucleotide Res.,* 7, 157, 1976.

170. **Steinberg, D. and Huttunen, J. K.,** The role of cyclic AMP in activation of hormone-sensitive lipase of adipose tissue, *Adv. Cyclic Nucleotide Res.,* 1, 47, 1972.

171. **Steinberg, D., Mayer, S. E., Khoo, J., Miller, E. A., Miller, R. E., Fredholm, B., and Eichner, R.,** Hormonal regulation of lipase, phosphorylase and glycogen synthase in adipose tissue, *Adv. Cyclic Nucleotide Res.,* 5, 549, 1975.

172. **Belfrage, P., Jergil, B., Strålfors, P., and Tornqvist, H.,** Hormone-sensitive lipase of rat adipose tissue: identification and some properties of the enzyme protein, *FEBS Lett.,* 75, 259, 1977.

173. **Belfrage, P., Jergil, B., Strålfors, P., and Tornqvist, H.,** Identification and some characteristics of the enzyme protein of the hormone-sensitive lipase from rat adipose tissue, in *Enzymes of Lipid Metabolism,* Gatt, S., Freysz, L., and Mandel, P., Eds., Plenum Press, New York, 1978, 113.

174. **Belfrage, P., Fredrikson, G., Nilsson, N. Ö., and Stralfors, P.,** Regulation of adipose tissue lipolysis: phosphorylation of hormone-sensitive lipase in intact rat adipocytes, *FEBS Lett.,* 111, 120, 1980.

175. **Fredrikson, G., Strålfors, P., Nilsson, N. Ö., and Belfrage, P.,** Hormone-sensitive lipase of rat adipose tissue. Purification and some properties, *J. Biol. Chem.,* 256, 6311, 1981.

176. **Strålfors, P. and Belfrage, P.,** Phosphorylation of hormone-sensitive lipase by cyclic AMP-dependent protein kinase, *J. Biol. Chem.,* 258, 15146, 1983.

177. **Strålfors, P., Björgell, P., and Belfrage, P.,** Hormonal regulation of hormone-sensitive lipase in intact adipocytes: identification of phosphorylated sites and effects on the phosphorylation by lipolytic hormones and insulin, *Proc. Natl. Acad. Sci. U.S.A.,* 81, 3317, 1984.

178. **Londos, C., Cooper, D. M. F., and Rodbell, M.,** Receptor-mediated stimulation and inhibition of adenylate cyclases: the fat cell as a model system, *Adv. Cyclic Nucleotide Res.,* 14, 163, 1981.

179. **Birnbaumer, L., Codina, J., Mattera, R., Cerione, R. A., Hildebrandt, J. D., Sunyer, T., Rojas, F. J., Caron, M. G., Lefkowitz, R. J., and Iyangar, R.,** Regulation of hormone receptors and adenylyl cyclases by guanine nucleotide binding N proteins, *Recent Progr. Horm. Res.,* 41, 41, 1985.

180. **Honnor, R. C., Dhillon, G. S., and Londos, C.,** cAMP-dependent protein kinase and lipolysis in rat adipocytes. II. Definition of steady-state relationship with lipolytic and antilipolytic modulators, *J. Biol. Chem.,* 260, 15130, 1985.

181. **Correze, C., Laudat, M. H., Laudat, P., and Nunez, J.,** Hormone-dependent lipolysis in fat-cells from thyroidectomized rats, *Mol. Cell. Endocrinol.,* 1, 309, 1974.

182. **Malbon, C. C., Moreno, F. J., Cabelli, R. J., and Fain, J. N.,** Fat cell adenylate cyclase and β-adrenergic receptors in altered thyroid states, *J. Biol. Chem.,* 253, 671, 1978.

183. **Ohisalo, J. J. and Stouffer, J. E.,** Adenosine, thyroid status and regulation of lipolysis, *Biochem. J.,* 178, 249, 1979.

184. **Goswami, A. and Rosenberg, I. N.,** Thyroid hormone modulation of epinephrine-induced lipolysis in rat adipocytes: a possible role of calcium, *Endocrinology (Baltimore),* 103, 2223, 1980.

185. **Malbon, C. C. and Graziano, M. P.,** Adenosine deaminase normalizes cyclic AMP responses of hypothyroid rat fat cells to forskolin, but not β-adrenergic agonists, *FEBS Lett.,* 155, 35, 1983.

186. **Chohan, P., Carpenter, C., and Saggerson, E. D.,** Changes in the anti-lipolytic action and binding to plasma membranes of N^6-L-phenylisopropyladenosine in adipocytes from starved and hypothyroid rats, *Biochem. J.,* 223, 53, 1984.

187. **Malbon, C. C., Graziano, M. P., and Johnson, G. L.,** Fat cell β-adrenergic receptor in the hypothyroid rat. Impaired interaction with the stimulatory component of adenylate cyclase, *J. Biol. Chem.,* 259, 3254, 1984.

188. **Malbon, C. C., Ropiejko, P. J., and Mangano, T. J.,** Fat cell adenylate cyclase system. Enhanced inhibition by adenosine and GTP in the hypothyroid rat, *J. Biol. Chem.,* 260, 2558, 1985.

189. **Saggerson, E. D.,** Sensitivity of adipocyte lipolysis to stimulatory and inhibitory agonists in hypothyroidism and starvation, *Biochem. J.,* in press, 1986.

190. **Shönhöfer, P., Skidmore, I. F., Bourne, H. R., Krishna, G. A., and Brodie, B. B.,** Influence of adrenal cortical hormones on the norepinephrine-induced lipolysis, *Arzneim. Forsch.,* 18, 1540, 1968.

191. **Allen, D. O. and Beck, R. R.,** Alterations in lipolysis, adenylate cyclase and adenosine 3′,5′-monophosphate levels in isolated fat cells following adrenalectomy, *Endocrinology (Baltimore),* 91, 504, 1972.

192. **Skidmore, I. F., Schönhöfer, P. S., Bourne, H. R., and Krishna, G.,** Effect of adrenalectomy and cortisone replacement on the lipolysis in fat tissue and fat cells, *Naunyn-Schmiedeberg's Arch. Pharmcol.,* 274, 113, 1972.

193. **Fernandez, B. M. and Saggerson, E. D.,** Alterations in response of rat white adipocytes to insulin, noradrenaline, corticotropin and glucagon after adrenalectomy, *Biochem. J.,* 174, 111, 1978.

194. **Saggerson, E. D.,** Increased antilipolytic effect of the adenosine "R-site" agonist N^6-(phenylisopropyl) adenosine in adipocytes from adrenalectomized rats, *FEBS Lett.,* 115, 127, 1980.

195. **Aitchison, R. E. D., Clegg, R. A., and Vernon, R. G.,** Lipolysis in rat adipocytes during pregnancy and lactation, *Biochem. J.,* 202, 243, 1982.

196. **Vernon, R. G., Finley, E., and Taylor, E.,** Adenosine and the control of lipolysis in rat adipocytes during pregnancy and lactation, *Biochem. J.,* 216, 121, 1983.
197. **Vernon, R. G. and Finley, E.,** Lipolysis in rat adipocytes during recovery from lactation. Response to noradrenaline and adenosine, *Biochem. J.,* 234, 229, 1986.
198. **Zapf, J., Waldvogel, M., and Froesch, E. R.,** Increased sensitivity of rat adipose tissue to the lipolytic action of epinephrine during fasting and its reversal during re-feeding, *FEBS Lett.,* 76, 135, 1977.
199. **Honnor, R. C. and Saggerson, E. D.,** Altered lipolytic response to glucagon and adenosine deaminase in adipocytes from starved rats, *Biochem. J.,* 188, 757, 1980.
200. **Dax, E. M., Partilla, J. S., and Gregerman, R. I.,** Increased sensitivity to epinephrine-stimulated lipolysis during starvation: tighter coupling of the adenylate cyclase complex, *Biochem. Biophys. Res. Commun.,* 101, 1186, 1981.
201. **Chohan, P. and Saggerson, E. D.,** Increased sensitivity of adipocyte adenylate cyclase to glucagon in the fasted state, *FEBS Lett.,* 146, 357, 1982.
202. **Zumstein, P., Zapf, J., Waldvogel, M., and Froesch, E. R.,** Increased sensitivity to lipolytic hormones of adenylate cyclase in fat cells of diabetic rats, *Eur. J. Biochem.,* 105, 187, 1980.
203. **Chatzipanteli, K. and Saggerson, E. D.,** Streptozotocin diabetes results in increased responsiveness of adipocyte lipolysis to glucagon, *FEBS Lett.,* 115, 135, 1983.
204. **Chatzipanteli, K. and Saggerson, E. D.,** Changes in the anti-lipolytic action and binding to plasma membranes of N^6-L-phenylisopropyl-adenosine in adipocytes from streptozotocin-diabetic rats, *Biochem. J.,* submitted for publication, 1986.
205. **Nilsson, N. Ö., Strålfors, P., Fredrikson, G., and Belfrage, P.,** Regulation of adipose tissue lipolysis: effects of noradrenaline and insulin on phosphorylation of hormone-sensitive lipase and on lipolysis in intact rat adipocytes, *FEBS Lett.,* 111, 125, 1980.
206. **Kather, H., Aktories, K., Schulz, G., and Jacobs, K. H.,** Islet activating protein discriminates the antipolytic mechanism of insulin from that of other antilipolytic compounds, *FEBS Lett.,* 161, 149, 1983.
207. **Butcher, R. W., Sneyd, J. G. T., Park, C. R., and Sutherland, E. W.,** Effect of insulin on adenosine 3′,5′-monophosphate in the rat epididymal fat pad, *J. Biol. Chem.,* 241, 1651, 1966.
208. **Jungas, R. L.,** Role of cyclic 3′,5′-AMP in the response of adipose tissue to insulin, *Proc. Natl. Acad. Sci. U.S.A.,* 56, 757, 1966.
209. **Butcher, R. W., Baird, C. E., and Sutherland, E. W.,** Effects of lipolytic and antilipolytic substances on adenosine 3′,5′-monophosphate levels in isolated fat cells, *J. Biol. Chem.,* 243, 1705, 1968.
210. **Manganiello, V. C., Murad, F., and Vaughan, M.,** Effects of lipolytic and antilipolytic agents on 3′,5′-adenosine monophosphate in fat cells, *J. Biol. Chem.,* 246, 2195, 1971.
211. **Fain, J. N. and Rosenberg, L.,** Antilipolytic action of insulin on fat cells, *Diabetes,* 21, 414, 1972.
212. **Khoo, J. C., Steinberg, D., Thompson, B., and Mayer, S. E.,** Hormonal regulation of adipocyte enzymes: the effects of epinephrine and insulin on the control of lipase, phosphorylase kinase, phosphorylase and glycogen synthase, *J. Biol. Chem.,* 248, 3823, 1973.
213. **Siddle, K. and Hales, C. N.,** The relationship between the concentration of adenosine 3′:5′-cyclic monophosphate and the antilipolytic action of insulin in isolated rat fat-cells, *Biochem. J.,* 142, 97, 1974.
214. **Londos, C., Honnor, R. C., and Dhillon, G. S.,** cAMP-dependent protein kinase and lipolysis in rat adipocytes. III. Multiple modes of insulin regulation of lipolysis and regulation of insulin responses by adenylate cyclase regulators, *J. Biol. Chem.,* 260, 15139, 1985.
215. **Rose, G. and Shapiro, B.,** Enzyme systems in adipose tissue participating in fatty acid esterification, *Bull. Res. Council Israel,* 9A, 15, 1960.
216. **Daniel, A. M. and Rubinstein, D.,** Fatty acid esterifying enzymes in rat adipose tissue homogenates, *Can. J. Biochem.,* 46, 1039, 1968.
217. **Jamdar, S. C., Shapiro, D., and Fallon, H. J.,** Triacylglycerol biosynthesis in the adipose tissue of the obese-hyperglycaemic mouse, *Biochem. J.,* 158, 327, 1976.
218. **Moller, F., Green, P., and Harkness, E. J.,** Soluble rat adipocyte phosphatidate phosphatase activity: characterization and effects of fasting and various lipids, *Biochim. Biophys. Acta,* 486, 359, 1977.
219. **Jamdar, S. C., Osborne, L. J., and Zeigler, J. A.,** Glycerolipid biosynthesis in rat adipose tissue. Influence of adipocyte size, *Biochem. J.,* 194, 193, 1981.
220. **Jamdar, S. C., Osborne, L. J., and Wells, G. N.,** Glycerolipid biosynthesis in rat adipose tissue. 12. Properties of Mg^{2+}-dependent and -independent phosphatidate phosphohydrolase, *Arch. Biochem. Biophys.,* 233, 370, 1984.
221. **Jamdar, S. C., Osborne, L. J., and Wells, G. N.,** Glycerolipid biosynthesis in rat adipose tissue. 13. Influence of age and cell size on substrate utilization, *Lipids,* in press, 1986.
222. **Coleman, R. A., Reed, B. C., Mackall, J. C., Student, A. K., Lane, M. D., and Bell, R. M.,** Selective changes in microsomal enzymes in triacylglycerol, phosphatidylcholine, and phosphatidylethanolamine biosynthesis during differentiation of 3T3-L1 preadipocytes, *J. Biol. Chem.,* 253, 7256, 1978.
223. **Grimaldi, P., Négrel, R., and Ailhaud, G.,** Induction of the triglyceride pathway enzymes and of lipolytic enzymes during differentiation in a 'preadipocyte' cell line, *Eur. J. Biochem.,* 84, 369, 1978.

224. **Roncari, D. A. K., Mack, E. Y. W., and Yip, D. K.,** Enhancement of microsomal phosphatidate phosphohydrolase and diacylglycerol acyltransferase activity by insulin during growth of rat adipocyte precursors in culture, *Can. J. Biochem.,* 57, 573, 1979.

225. **Lawson, N., Pollard, A. D., Jennings, R. J., Gurr, M. I., and Brindley, D. N.,** The activities of lipoprotein lipase and of enzymes involved in triacylglycerol synthesis in rat adipose tissue, *Biochem. J.,* 200, 285, 1981.

226. **Moller, F. and Hough, M. R.,** Effect of salts on membrane binding and activity of adipocyte phosphatidate phosphohydrolase, *Biochim. Biophys. Acta,* 711, 521, 1982.

227. **Lamb, R. G. and Fallon, H. J.,** An enzymatic explanation for increased hepatic triglyceride formation in rats fed high sugar diets, *J. Clin. Invest.,* 51, 53a, 1972.

228. **Sturton, R. G. and Brindley, D. N.,** Problems encountered in measuring the activity of phosphatidate phosphohydrolase, *Biochem. J.,* 171, 263, 1978.

229. **Sturton, R. G. and Brindley, D. N.,** Factors controlling the metabolism of phosphatidate by phosphohydrolase and phospholipase A-type activities, *Biochim. Biophys. Acta,* 619, 494, 1980.

230. **Jamdar, S. C. and Osborne, L. J.,** Glycerolipid biosynthesis in rat adipose tissue. 11. Effects of polyamines on Mg^{2+}-dependent phosphatidate phosphohydrolase, *Biochim. Biophys. Acta,* 752, 79, 1983.

231. **Butterwith, S. C., Martin, A., and Brindley, D. N.,** Can phosphorylation of phosphatidate phosphohydrolase by a cyclic AMP-dependent mechanism regulate its activity and subcellular distribution and control hepatic glycerolipid synthesis?, *Biochem. J.,* 222, 487, 1984.

232. **Saggerson, E. D. and Rider, M. H.,** Fatty acids and the control of adipocyte glycerolphosphate acyltransferase by noradrenaline, *Biochem. J.,* 231, 495, 1985.

233. **Lederer, B. and Hers, H-G.,** A kinetic study of glycerophosphate acyltransferase of rat adipocytes in relation to its control by noradrenaline, *Biochem. J.,* 226, 269, 1985.

234. **Kirsch, D., Obermaier, B., and Häring, H. U.,** Phorbol esters enhance basal D-glucose transport but inhibit insulin stimulation of D-glucose transport and insulin binding in isolated rat adipocytes, *Biochem. Biophys. Res. Commun.,* 128, 824, 1985.

235. **Skoglund, G., Hansson, A., and Ingelman-Sundberg, M.,** Rapid effects of phorbol esters on isolated rat adipocytes — relationship to the action of protein kinase C, *Eur. J. Biochem.,* 148, 407, 1985.

236. **Van de Werve, G., Proietto, J., and Jeanrenaud, B.,** Tumour-promoting phorbol esters increase basal and inhibit insulin-stimulated lipogenesis in rat adipocytes without decreasing insulin binding, *Biochem. J.,* 225, 523, 1985.

237. **Nishizuka, Y.,** The role of protein kinase C in cell surface signal transduction and tumour promotion, *Nature (London),* 308, 693, 1984.

238. **Castagna, M., Takai, Y., Kaibuchi, K., Sano, K., Kikkawa, U., and Nishizuka, Y.,** Direct activation of calcium-activated, phospholipid-dependent protein kinase by tumour-promoting phorbol esters, *J. Biol. Chem.,* 257, 7847, 1982.

239. **Inoue, M., Kishimoto, A., Takai, Y., and Nishizuka, Y.,** Studies on a cyclic nucleotide-independent protein kinase and its proenzyme in mammalian tissues, *J. Biol. Chem.,* 252, 7610, 1977.

240. **Kuo, J. F., Andersson, R. G. G., Wise, B. C., Mackerlova, L., Salomonsson, I., Brackett, N. L., Katoh, N., Shoji, M., and Wrenn, R. W.,** Calcium-dependent protein kinase: widespread occurrence in various tissues and phyla of the animal kingdom and comparison of effects of phospholipid, calmodulin and trifluoroperazine, *Proc. Natl. Acad. Sci. U.S.A.,* 77, 7039, 1980.

241. **Minakuchi, R., Takai, Y., Yu, B., and Nishizuka, Y.,** Widespread occurrence of calcium-activated, phospholipid-dependent protein kinase in mammalian tissues, *J. Biochem. (Tokyo),* 89, 1651, 1981.

242. **Garrison, J. C., Johnsen, D. E., and Campanile, C. P.,** Evidence for the role of phosphorylase kinase, protein kinase C, and other Ca^{2+}-sensitive protein kinases in the response of hepatocytes to angiotensin II and vasopressin, *J. Biol. Chem.,* 259, 3283, 1984.

243. **Pollard, A. D. and Brindley, D. N.,** Effects of vasopressin and corticosterone on fatty acid metabolism and on the activities of glycerolphosphate acyltransferase and phosphatidate phosphohydrolase in rat hepatocytes, *Biochem. J.,* 217, 461, 1984.

244. **Hall, M., Taylor, S. J., and Saggerson, E. D.,** Persistent activity modification of phosphatidate phosphohydrolase and fatty acyl-CoA synthetase on incubation of adipocytes with the tumour promoter 12-O-tetradecanoylphorbol 13-acetate, *FEBS Lett.,* 179, 351, 1985.

245. **Lawson, N., Jennings, R. J., Pollard, A. D., Sturton, R. G., Ralph, S. J., Marsden, C. A., Fears, R., and Brindley, D. N.,** Effects of chronic modification of dietary fat and carbohydrate in rats. The activities of some enzymes of hepatic glycerolipid synthesis and the effects of corticotropin injection, *Biochem. J.,* 200, 265, 1981.

246. **Vavrecka, M., Mitchell, M. P., and Hubscher, G.,** The effect of starvation on the incorporation of palmitate into glycerides and phospholipids of rat liver homogenates, *Biochem. J.,* 115, 139, 1969.

247. **Murthy, V. K. and Shipp, J. C.,** Synthesis and accumulation of triglycerides in liver of diabetic rats. Effects of insulin treatment, *Diabetes,* 28, 472, 1979.

248. **Whiting, P. H., Bowley, M., Sturton, R. G., Pritchard, P. H., Brindley, D. N., and Hawthorne, J. N.,** The effect of chronic diabetes, induced by streptozotocin, on the activities of some enzymes of glycerolipid synthesis in rat liver, *Biochem. J.,* 168, 147, 1977.

249. **Woods, J. A., Knauer, T. E., and Lamb, R. G.,** The acute effects of streptozotocin-induced diabetes on rat liver glycerolipid biosynthesis, *Biochim. Biophys. Acta,* 666, 482, 1982.

250. **Lawson, N., Jennings, R. J., Fears, R., and Brindley, D. N.,** Antagonistic effects of insulin on the corticosterone-induced increase of phosphatidate phosphohydrolase activity in isolated rat hepatocytes, *FEBS Lett.,* 143, 9, 1982.

251. **Lawson, N., Pollard, A. D., Jennings, R. J., and Brindley, D. N.,** Effects of corticosterone and insulin on enzymes of triacylglycerol synthesis in isolated rat hepatocyes, *FEBS Lett.,* 146, 204, 1982.

252. **Pittner, R. A., Fears, R., and Brindley, D. N.,** Effects of cyclic AMP, glucocorticoids and insulin on the activities of phosphatidate phosphohydrolase, tyrosine aminotransferase and glycerol kinase in isolated rat hepatocytes in relation to the control of triacylglycerol synthesis and gluconeogenesis, *Biochem. J.,* 225, 455, 1985.

253. **Pittner, R. A., Fears, R., and Brindley, D. N.,** Interactions of insulin, glucagon and dexamethasone in controlling the activity of glycerol phosphate acyltransferase and the activity and subcellular distribution of phosphatidate phosphohydrolase in cultured rat hepatocytes, *Biochem. J.,* 230, 525, 1985.

254. **Pritchard, P. H., Bowley, M., Burditt, S. L., Cooling, J., Glenny, H. P., Lawson, N., Sturton, R. G., and Brindley, D. N.,** The effects of acute ethanol feeding and of chronic benfluorex administration on the activities of some enzymes of glycerolipid synthesis in rat liver and adipose tissue, *Biochem. J.,* 166, 639, 1977.

255. **Jamdar, S. C. and Osborne, L. J.,** Glycerolipid biosynthesis in rat adipose tissue. 7. Effect of age, site of adipose tissue and cell size, *Biochim. Biophys. Acta,* 665, 145, 1981.

256. **Belfiore, F., Rabuazzo, A. M., Borzi, V., and Iannello, S.,** Triglyceride synthesis in the adipose tissue of obese patients. A study of the key enzyme phosphatidate phosphatase, *Diabetologia,* 15, 218, 1978.

257. **Wilson, J. E.,** Ambiquitous enzymes: variation in intracellular distribution as a regulatory mechanism, *Trends Biochem. Sci.,* 3, 124, 1978.

258. **Brindley, D. N.,** Intracellular translocation of phosphatidate phosphohydrolase and its possible role in the control of glycerolipid synthesis, *Progr. Lipid Res.,* 23, 115, 1984.

259. **Martin-Sanz, P., Hopewell, R., and Brindley, D. N.,** Long-chain fatty acids and their acyl-CoA esters cause the translocation of phosphatidate phosphohydrolase from the cytosolic to the microsomal fraction of rat liver, *FEBS Lett.,* 175, 284, 1984.

260. **Martin-Sanz, P., Hopewell, R., and Brindley, D. N.,** Spermine promotes the translocation of phosphatidate phosphohydrolase from the cytosol to the microsomal fraction of rat liver and it enhances the effects of oleate in this respect, *FEBS Lett.,* 179, 262, 1985.

261. **Hopewell, R., Martin-Sanz, P., Martin, A., Saxton, J., and Brindley, D. N.,** Regulation of the translocation of phosphatidate phosphohydrolase between the cytosol and the endoplasmic reticulum of rat liver. Effects of unsaturated fatty acids, spermine, nucleotides, albumin and chlorpromazine, *Biochem. J.,* 232, 485, 1985.

262. **Cascales, C., Mangiapane, E. H., and Brindley, D. N.,** Oleic acid promotes the activation and translocation of phosphatidate phosphohydrolase from the cytosol to particulate fractions of isolated rat hepatocytes, *Biochem. J.,* 219, 911, 1984.

263. **Angel, A., Desai, K. S., and Halperin, M. L.,** Intracellular accumulation of free fatty acids in isolated white adipose cells, *J. Lipid Res.,* 12, 104, 1971.

264. **Angel, A., Desai, K. S., and Halperin, M. L.,** Reduction in adipocyte ATP by lipolytic agents: relation to intracellular free fatty acid accumulation, *J. Lipid Res.,* 12, 203, 1971.

265. **Heindel, J. J., Cushman, S. W., and Jeanrenaud, B.,** Cell associated fatty acid levels and energy-requiring processes in mouse adipocytes, *Am. J. Physiol.,* 226, 16, 1974.

266. **Baht, H. S. and Saggerson, D.,** unpublished work.

267. **Taylor, S. J. and Saggerson, D.,** unpublished work.

268. **Cheng, C. H. K. and Saggerson, D.,** unpublished work.

INDEX

Milton Keynes UK
Ingram Content Group UK Ltd.
UKHW051935141024
449569UK00027B/1499